Applied Adaptive Statistical Methods

ASA-SIAM Series on Statistics and Applied Probability

The ASA-SIAM Series on Statistics and Applied Probability is published jointly by the American Statistical Association and the Society for Industrial and Applied Mathematics. The series consists of a broad spectrum of books on topics in statistics and applied probability. The purpose of the series is to provide inexpensive, quality publications of interest to the intersecting membership of the two societies.

O'Gorman, T. W., *Applied Adaptive Statistical Methods: Tests of Significance and Confidence Intervals*

Ross, T. J., Booker, J. M., and Parkinson, W. J., eds., *Fuzzy Logic and Probability Applications: Bridging the Gap*

Nelson, W. B., *Recurrent Events Data Analysis for Product Repairs, Disease Recurrences, and Other Applications*

Mason, R. L. and Young, J. C., *Multivariate Statistical Process Control with Industrial Applications*

Smith, P. L., *A Primer for Sampling Solids, Liquids, and Gases: Based on the Seven Sampling Errors of Pierre Gy*

Meyer, M. A. and Booker, J. M., *Eliciting and Analyzing Expert Judgment: A Practical Guide*

Latouche, G. and Ramaswami, V., *Introduction to Matrix Analytic Methods in Stochastic Modeling*

Peck, R., Haugh, L., and Goodman, A., *Statistical Case Studies: A Collaboration Between Academe and Industry, Student Edition*

Peck, R., Haugh, L., and Goodman, A., *Statistical Case Studies: A Collaboration Between Academe and Industry*

Barlow, R., *Engineering Reliability*

Czitrom, V. and Spagon, P. D., *Statistical Case Studies for Industrial Process Improvement*

Applied Adaptive Statistical Methods
Tests of Significance and Confidence Intervals

Thomas W. O'Gorman
Northern Illinois University
DeKalb, Illinois

siam
Society for Industrial and Applied Mathematics
Philadelphia, Pennsylvania

ASA
American Statistical Association
Alexandria, Virginia

10 9 8 7 6 5 4 3 2 1

Library of Congress Cataloging-in-Publication Data

O'Gorman, Thomas W.
Applied adaptive statistical methods : tests of significance and confidence intervals / Thomas W. O'Gorman.
p. cm. – (ASA-SIAM series on statistics and applied probability)
Includes bibliographical references and index.
ISBN 0-89871-553-9 (pbk.)
1. Adaptive sampling (Statistics). 2. Estimation theory. 3. Distribution (Probability theory). 4. Mathematical statistics. I.Title. II. Series.

QA276.6 O35 2004
519.5'2–dc22

2003061294

To my niece, LeAnne Dunn;
my wife, Martha;
and my children, Kelly and Tim.

Contents

Preface

Statistical procedures that are modified based on a preliminary inspection of the data are called adaptive statistical methods. The adaptive methods described in this book can be used to produce adaptive tests of significance, adaptive confidence intervals, and adaptive estimates. The first practical adaptive tests were proposed by Randles and Hogg (1973) and by Hogg, Fisher, and Randles (1975). Researchers were interested in adaptive tests of significance because they had the potential of having higher power than the traditional tests when the measurements were not normally distributed. However, since these early adaptive methods were based on ranks, their application was somewhat limited. In contrast, modern adaptive methods not based on ranks can be used with a wide variety of statistical models. For many real-world situations modern adaptive tests of significance are more powerful than traditional tests of significance and adaptive confidence intervals are often narrower than traditional confidence intervals.

In the past, doubts about the integrity of adaptive procedures may have discouraged their use. I believe that many researchers felt that there was something improper about using the data itself to adjust the test procedure. Although it could be demonstrated that rank-based adaptive tests would maintain their significance level, there may have been some concern about their validity. Another obstacle to the use of adaptive methods was the lack of available software. Although the first adaptive procedures were relatively simple to perform, they required software in order to be practical. Consequently, adaptive methods have not been widely used for the analysis of observational and experimental data.

In this book I hope to overcome many of the obstacles to the use of adaptive methods. Modern adaptive methods are more general than the earlier methods and sufficient software has been developed to make adaptive tests easy to use for many real-world problems. Also, the wide availability of fast computers has made it possible to use computationally intensive methods to ensure that these adaptive methods are fair. Every adaptive test of significance described in this book has an actual level of significance that is close to the nominal value. In addition, the adaptive confidence intervals have actual coverage probabilities that are close to their nominal confidence levels.

This book is intended to serve as an introduction and guide to many of the practical adaptive statistical methods that have been developed over the last 30 years. I will concentrate on the more modern adaptive methods that have been developed in the last few years because these methods tend to be more general than the older, rank-based methods. Adaptive methods will be presented for one-sample and two-sample tests, for one-way layouts, and for interaction effects in two-way designs. In addition, an adaptive test for a subset of coefficients in a linear model will be described. I will also present an adaptive approach to the construction of confidence intervals and an adaptive estimator.

I believe that these modern adaptive methods are reasonably easy to understand and use. In each chapter I have tried to present the material at a level so that someone with a knowledge of multiple regression analysis should be able to understand the adaptive procedures. Some matrix algebra is used to describe the adaptive weighting methods.

The book is arranged so that the basic adaptive test procedures are explained in the first few chapters. The rank-based test of Hogg, Fisher, and Randles (1975) is explained in the first chapter. An adaptive weighted least squares (WLS) test procedure is explained in Chapter 2. This is the most important material in the book since it contains most of the key ideas of modern adaptive testing. I have kept this material at a level so that readers with a limited mathematical background should be able to understand the adaptive weighting procedure. No matrix algebra is used in Chapter 2 and the permutation method is easy to understand when used with two-sample tests. Since some readers will be understandably skeptical about the validity of the adaptive test, I demonstrate it through simulation studies. I also demonstrate in Chapter 2 that the power of the adaptive test is often greater than the power of the traditional two-sample t test.

A general approach to adaptive testing is presented in Chapter 3. This approach to the testing of a subset of coefficients in a linear model is essential for understanding the test procedures used throughout the book. Readers who have had a course in regression analysis should be able to follow the presentation without difficulty. The permutation methods used with these multiple regression models are probably new to most readers, but they are fully explained in this chapter.

In Chapter 4 I describe the software that is necessary to perform the adaptive WLS test. This software is included in Appendix A of this book. In the first section of Chapter 4 the correct use of this software is illustrated for some simple models. In each of the remaining sections of that chapter, I present examples of adaptive test procedures. These examples include the adaptive analysis of data from a one-way layout, an adaptive test of the slope in a simple linear regression, an adaptive test for slope in a multiple linear regression, an adaptive analysis of data from a two-way layout, and an adaptive test for the treatment effect in a multicenter clinical trial. In each section an example data set is fully analyzed using the SAS macros and the results of a simulation study are shown to demonstrate that the adaptive test can be more powerful than the traditional test when the error distribution is skewed or has long tails.

Recognizing that many researchers have paired data, an adaptive test for paired data is included in Chapter 5. With nonnormal error distributions this test often has greater power than the paired t test. The adaptive testing methods can also be used to construct the adaptive confidence intervals described in Chapter 6. Some of the calculations are more difficult in this area, but I have avoided computational details in order to focus on the proper use of the adaptive methods. SAS macros, which are included in Appendix A, have also been developed to compute the adaptive confidence intervals.

In Chapter 7 an adaptive estimator is proposed for the regression parameters in a linear model. The adaptive estimator is a type of robust estimator that is often more precise than the traditional estimator for nonnormal error models.

Chapter 8 contains several important adaptive methods. An adaptive multiple comparison procedure is described and evaluated. I also discuss the problems that can arise when the shift model does not hold. If the shift model is not appropriate, a test of location and scale may be used. Both adaptive and nonadaptive versions of these tests are described in this chapter. In addition, I show that the adaptive tests are not greatly influenced by the presence of a large number of ties in the data.

The results of the simulation studies throughout the book show that the adaptive methods are often more powerful than traditional methods. However, the research has also shown that adaptive methods are not suitable for the analysis of every data set. There appears to be no advantage to using adaptive methods for small data sets with fewer than ten observations. Also, I am not aware of any adaptive approach to logistic regression models or random effects models.

My interest in adaptive methods grew out of research into epidemiological methods at the University of Iowa. At that time I was a member of a research team, led by Robert (Skip) Woolson, that was looking for more effective ways of selecting variables in case-control studies. I had thought that a more effective two-sample test procedure might increase the efficiency of variable selection methods in case-control studies and had pursued the development of some rank-based adaptive methods for two-sample tests. Several years later, realizing that adaptive methods would be much more valuable if they could be employed in a variety of circumstances, I attempted to develop an adaptive approach to tests for a subset of regression coefficients. This adaptive method has turned out to be a fruitful line of research and has formed the basis of this book.

I would like to thank my colleague Alan Polansky for encouraging me to write this book and for the suggestions he made, especially with regard to the smoothing methods that are used. My editor at SIAM, Linda Thiel, was constantly supportive and encouraging. I am also indebted to Bob Hogg, Ulric Lund, Herbert Büning, and Bob Rodriguez for their valuable suggestions, which have improved the quality of the manuscript.

Chapter 1

Introduction

Adaptive statistical methods use the data to determine which statistical procedure is most appropriate. Adaptive methods are usually performed in two steps. In the first step a selection statistic is computed that estimates the shape of the error distribution. In the second step the selection statistic is used to determine an effective statistical procedure for the error distribution. The advantage of adaptive methods is that, for tests of significance, the adaptive approach can increase the power of the test if the error distributions are skewed. The adaptive approach can make confidence intervals narrower than is traditionally the case.

Despite the fact that a practical and effective adaptive testing method was published many years ago by Hogg, Fisher, and Randles (1975), it is rarely used to analyze data. Some researchers may be reluctant to use adaptive methods because they have a mistaken belief that the methods may not be sound. Another reason that adaptive methods are not widely used is that, until the last few years, there has not been a general adaptive approach to tests of significance and confidence intervals that could easily be applied in practice. Another problem has been the lack of suitable software.

In this book we will encourage the use of adaptive methods by presenting evidence that they can be valid and effective if constructed properly. We will also describe a general approach to the construction of adaptive tests and confidence intervals and will provide some software to enable researchers to carry out these procedures. Unlike earlier adaptive approaches, which were often limited to the analysis of data from two independent samples and from one-way layouts, the general adaptive test described in this book can usually be used instead of the traditional F test to perform a test of significance on any subset of regression coefficients in a linear model. Thus, this adaptive test can be used for basic tests of significance and for more complex tests that may involve testing several slopes in a multiple linear regression.

Recently, there have been advances in the construction of adaptive confidence intervals. With nonnormal data these confidence intervals are often narrower than traditional confidence intervals. Because adaptive confidence intervals are derived from the adaptive test procedures, their properties are consistent with the results for the corresponding test of significance. Also, because the same general adaptive approach is used for tests and confidence intervals, little extra effort is required to understand the methods needed to produce adaptive confidence intervals.

Adaptive methods can also be used to estimate the parameters in a linear model. The same weighting scheme that is used in adaptive tests can be used in a linear model to downweight outliers. These adaptive estimators have been shown to compare favorably to ordinary least squares (OLS) estimators and to some robust estimators.

In order to facilitate the calculation of the test statistics and the confidence intervals, computer software has been developed for most of the methods described in this book. Because the adaptive test procedure is fairly straightforward, it is relatively easy to use the output from existing software in an adaptive weighting scheme. The software that the author has developed is in the form of SAS macros, which are given in Appendix A. These macros, along with some documentation, can also be downloaded from the author's Web site (*www.math.niu.edu/~ogorman/*).

1.1 Some Advantages of Adaptive Methods

Many devices used in our modern world are adaptive in nature. Many cameras adjust for lighting conditions automatically. Many clothes dryers adjust the amount of drying time depending on the level of humidity in the clothes. Yet in statistics we often use a fixed, or nonadaptive, approach to the analysis of data, even when we suspect that the usual inference assumptions may not be met.

There are a variety of reasons to explain the lack of flexibility in traditional statistical methods. First, we know from simulation studies that most of the traditional tests and confidence intervals are robust, at least as far as validity is concerned. That is, the traditional tests maintain their actual significance level close to their nominal significance level, and the traditional confidence intervals maintain their coverage probabilities close to their nominal coverage probabilities. Because their significance level and coverage probabilities are relatively unaffected by violations of the assumptions, we say that these methods are *robust for validity*. Also, researchers continue to use traditional nonadaptive methods because they are available in statistical software packages and are fairly easy to explain.

However, traditional methods have a serious defect. If the distribution of errors is nonnormal, the power of traditional tests can be much less than the power of adaptive tests, and traditional confidence intervals can be wider than adaptive confidence intervals. We say that these traditional methods are not *robust for efficiency*. The advantage of adaptive tests, which do maintain their significance level and hence are robust for validity, is that they are often more powerful than traditional tests. When adaptive confidence intervals are constructed they are often narrower than the traditional intervals. The power advantage of adaptive methods is important if much time and effort has been expended to collect the data.

Another advantage of adaptive methods is that they automatically downweight outliers, which has the effect of making the results less sensitive to observations that do not agree with the model. The robust nature of adaptive methods makes them ideally suited to the analysis of complex data. With adaptive methods we are able to modify traditional methods so that they are as robust as many nonparametric methods.

1.2 A Brief History of Adaptive Methods

Many adaptive one-sample and two-sample tests of significance have been published over the last 30 years. The first practical adaptive procedure for testing the null hypothesis that the center of a symmetric distribution equals a specified value was proposed by Randles and Hogg (1973). In that paper they also proposed a two-sample adaptive procedure for testing the null hypothesis that two symmetric distributions are identical. A few years later Hogg, Fisher, and Randles (1975) proposed a two-sample adaptive test that did not require symmetric error distributions. This two-sample test, which will be called the HFR test, used selection statistics that were measures of asymmetry and tailweight to select one of several rank tests. The HFR test played an important role in the development of rank-based adaptive tests because it was easy to compute and was the first practical two-sample test that was robust for validity and efficiency. Several years later Hüsler (1987) proposed an interesting modification of the two-sample HFR test that had high power but did not always maintain the nominal level of significance. Hill, Padmanabhan, and Puri (1988) proposed adaptive methods for performing multiple comparisons and for testing ordered alternatives. Although the approach used in the HFR test has been extended to the analysis of data from the one-way layout by Büning (1996), it has not been extended to tests that could be used with regression data or multiway layouts.

In their work, Hogg, Fisher, and Randles (1975) used measures of asymmetry and tailweight as selection statistics in the adaptive test. If the selection statistics fell into one of the regions defined by the adaptive procedure, then a certain set of rank scores was selected, whereas if the selection statistic fell into a different region, then a different set of rank scores would be used in the test. One problem with their approach was that, if the selection statistics were near a border of a region defined by the adaptive procedure, a small change in the data could create a small change in the selection statistics, which could cause a large change in the rank scores. To get around this difficulty, several researchers have attempted to use continuously adaptive selection statistics. One continuously adaptive test was proposed by Ruberg (1986) for the two-sample problem. Unlike the HFR test, this continuously adaptive rank-based method has the advantage of changing only slightly when small changes are made to the data. Unfortunately, the test is rather complex and O'Gorman (1996) has shown that it can have relatively low power in certain situations. A continuously adaptive rank-based test was proposed for the one-way layout by O'Gorman (1997a), which had relatively good power compared to the F test and the Kruskal–Wallis test.

Although Randles and Hogg (1973) appeared to have the earliest practical adaptive tests, some theoretical work had also been done on the adaptive testing problem by Hájek (1962), who suggested a method to estimate the best score function. It does not appear that any practical procedure has been developed based on that approach. Another adaptive approach was proposed by Behnen and Neuhaus (1989) to determine the linear rank scores. They emphasized the importance of considering models that are more general than the shift model.

Adaptive tests have also been proposed for the two-sample scale problem and the combined test for location and scale. Hall and Padmanabhan (1997) proposed three adaptive tests for the two-sample scale problem. Büning and Thadewald (2000) proposed an adaptive test that is sensitive to differences in location or scale. This test is described in Chapter 8.

One problem with rank-based adaptive tests is that they cannot easily be generalized to testing parameters in linear models. To overcome this limitation a more general approach to adaptive testing was proposed by O'Gorman (2001a). In this adaptive procedure the tests of significance were formulated as tests for a subset of regression coefficients in a linear model. Consequently, with this adaptive test we can easily perform most of the commonly used tests of significance, including the two-sample test, a test for a slope in a linear regression, and a test for interaction in a two-way factorial design. The test proposed by O'Gorman (2001a) represented a departure from the previous adaptive tests because it was not based on ranks. Instead, a weighting function was used to weight observations, after which the weighted least squares (WLS) method was used to fit the weighted data. Permutation methods were then used to perform the test in a manner that ensured that the adaptive test would maintain the level of significance. An improved version of this test was published by O'Gorman (2002). The improved version is easier to understand and slightly more powerful than the previous version. Many of the tests and confidence intervals described in this book are derived from this adaptive test procedure.

In contrast to the large literature on adaptive testing, there have been relatively few articles on adaptive confidence intervals. Clarke (1994) evaluated the performance of several adaptive confidence intervals for the mean of a symmetric distribution and O'Gorman (2001c) proposed an adaptive confidence interval for a single regression parameter in a linear model. The effectiveness of O'Gorman's adaptive confidence interval procedure was evaluated for the slope in a simple linear regression and for a confidence interval for the difference between two means. O'Gorman (2001c) showed, for nonnormal error distributions, that an adaptive confidence interval could be narrower, on average, than the nonadaptive confidence interval, while it maintained the coverage probability at the nominal value.

Several authors have suggested adaptive methods to estimate parameters. Hogg and Lenth (1984) proposed an adaptive estimator of the mean of a symmetric distribution. They used a selection statistic to determine if a mean, a 25% trimmed mean, or a median should be used as an estimate of the population mean. Hill and Padmanabhan (1991) proposed an adaptive estimator for the one-sample problem. Han and Hawkins (1994) also proposed an adaptive estimator of the mean. Two estimators of the slope of a regression line that use an adaptive M-regression approach were proposed by Yuh and Hogg (1988), and an adaptive combination of pairs of linear regression estimation methods was proposed by Dodge and Jurečková (2000). Using a continuously adaptive approach, O'Gorman (2001b) proposed an estimator for slope parameters in a linear model and also proposed an estimator of the mean of a symmetric distribution. In addition to these adaptive methods, there are many nonadaptive robust methods of estimation that have the same objectives as the adaptive methods. The relationship between these robust methods and adaptive methods will be discussed in section 1.5.

Although this book is primarily concerned with tests of significance and confidence intervals, it should be noted that adaptive methods are used in several other areas within statistics. Adaptive designs can be used with experimental procedures in order to increase the chance of rejecting the null hypothesis. In some of these adaptive designs the experiment is divided into two trials and a decision is made, after the data from the first trial are analyzed, to reject the null hypothesis, to stop the trial without rejecting the null hypothesis, or to continue to the second stage of the trial. For an example of adaptive design see Liu and Chi (2001). The adaptive approach is also used in the analysis of time-series data with adaptive filters. In the field of sampling, adaptive cluster-sampling techniques have been proposed.

Although there appears to be considerable growth in the application of adaptive methods to a wide variety of statistical problems, this book will not cover adaptive sampling, adaptive filters, or adaptive designs. We will concentrate on adaptive versions of the most commonly used statistical methods. That is, our objective in this book is to fully describe an adaptive approach to most of the tests and confidence intervals found in most elementary introductory statistics texts.

1.3 An Adaptive Test Based on Ranks

1.3.1 The Role of Rank Tests

The adaptive test that has received the most attention is the rank-based two-sample adaptive test proposed by Hogg, Fisher, and Randles (1975). The HFR test is interesting because it illustrates a simple and effective way of developing an adaptive test that maintains its level of significance. The essential features of this test will be described in this section; an extension and modification of this rank-based test that was proposed by Büning (1996) will be described in Chapter 8.

For this two-sample adaptive test we let $X_1, \ldots, X_m$ be a random sample from population 1 and $Y_1, \ldots, Y_n$ be a random sample taken independently from population 2. We assume that the random variables in population 1 have a cumulative distribution function (c.d.f.) of $F(x)$ and that the c.d.f. of the random variables in population 2 is shifted to the right or left of $F(x)$ but is otherwise identical. Thus, we assume that the random variables in population 2 have a c.d.f. of $F(x - \Delta)$, where Δ is the shift parameter. In this section we will test $H_0 : \Delta = 0$ against the one-sided alternative $H_a : \Delta > 0$.

We begin the discussion of the HFR test by observing that some rank tests work better than others in certain circumstances. For example, the Wilcoxon rank-sum test is the locally most powerful rank test for detecting a shift in a logistic distribution, whereas the median test is effective in detecting a shift in a double-exponential distribution. We could make an adaptive test by establishing some rule whereby we would use the median test if the empirical distribution appeared to have long tails; otherwise we would use the Wilcoxon test. Would such a test be fair? That is, would the actual Type I error be anywhere near the nominal level of α?

Such a test would maintain its stated level of significance only if it were carefully constructed. In this hypothetical adaptive test, both test statistics are based on the ranks of the combined data. As we shall demonstrate in the next section, if the test statistics used in the adaptive test are based on ranks, then it is easy to construct a valid adaptive test.

1.3.2 The HFR Test

We now proceed to describe an adaptive two-sample test proposed by Hogg, Fisher, and Randles (1975). In the HFR test, the choice of the most appropriate test is based on two selection statistics obtained by combining the $N = m + n$ observations in the samples. Let $\bar{U}_{.05}$ be the average of the $.05N$ largest observations in the combined sample, let $\bar{L}_{.05}$ be the average of the $.05N$ smallest observations, and let $\bar{M}_{.50}$ be the average of the $.5N$ middle observations. To construct an adaptive test we will use the selection statistics to determine the rank test to be used. The first selection statistic, which measures the asymmetry in the sample distribution, is

$$Q_3 = \frac{\bar{U}_{.05} - \bar{M}_{.5}}{\bar{M}_{.5} - \bar{L}_{.05}}.$$

If the empirical distribution is nearly symmetric, then Q_3 should approximate 1.0. If the empirical distribution is skewed to the right, then Q_3 may greatly exceed 1.0. The second selection statistic measures the distance between the upper and lower tails of the sample distribution. It is defined as

$$Q_4 = \frac{\bar{U}_{.05} - \bar{L}_{.05}}{\bar{U}_{.5} - \bar{L}_{.5}},$$

where $\bar{U}_{.5}$ is the average of the largest $.5N$ observations in the combined sample and $\bar{L}_{.5}$ is the average of the smallest $.5N$ observations. For the uniform distribution Q_4 is equal to 1.9, but for long-tailed distributions Q_4 may exceed 3. In the HFR test Q_3 and Q_4 are the selection statistics that are used to select the most appropriate rank test.

The adaptive HFR test uses two-sample linear rank tests as component tests in the overall test. In order to calculate the test statistic, we first combine the two samples and find the ranks of the observations in both samples. Often, in nonparametric statistics, we use the ranks themselves in the test statistic, as we do with the Wilcoxon rank-sum test. However, in some nonparametric tests we use some function of the ranks instead of the ranks themselves. The function of the rank is called the rank score, which will be denoted by $a(R_i)$ for an observation whose rank is R_i. Let $\{R_1, \ldots, R_n\}$ be the ranks of the observations in the second sample and let $\{a(R_j),\ j = 1, \ldots, n\}$ be the set of scores for those ranks. Two-sample rank statistics are based on $S = \sum_{j=1}^{n} a(R_j)$, which is the sum of the rank scores for the observations in the second sample.

The Wilcoxon test is a popular nonparametric test that is effective in detecting a shift in a logistic distribution. Consequently, Hogg, Fisher, and Randles (1975) proposed that if the distribution is not too highly skewed and has moderate tails, then the Wilcoxon rank-sum test will be used. In the Wilcoxon test, the scores are simply the ranks, so that

$$a_W(i) = i \quad \text{if } 1 \le i \le N,$$

and it follows that S equals the sum of the ranks in the second sample.

The median test is a two-sample linear rank test based on a count of the number of observations in the second sample that equal or exceed the median. The median test is effective in detecting a shift if the observations appear to be from a heavy-tailed distribution. This test statistic can be expressed as a linear rank statistic by using the scores

$$a_M(i) = \begin{cases} 1 & \text{if } i \ge (N+1)/2, \\ 0 & \text{otherwise.} \end{cases}$$

Other tests can be constructed for special circumstances. If the observations appear to be from a light-tailed distribution, we could give relatively larger weights to the largest and smallest ranks by using the following scores in a linear rank test statistic:

$$a_{LT}(i) = \begin{cases} i - (N+1)/4 & \text{if } i < (N+1)/4, \\ 0 & \text{if } (N+1)/4 \le i \le 3(N+1)/4, \\ i - 3(N+1)/4 & \text{if } i > 3(N+1)/4. \end{cases}$$

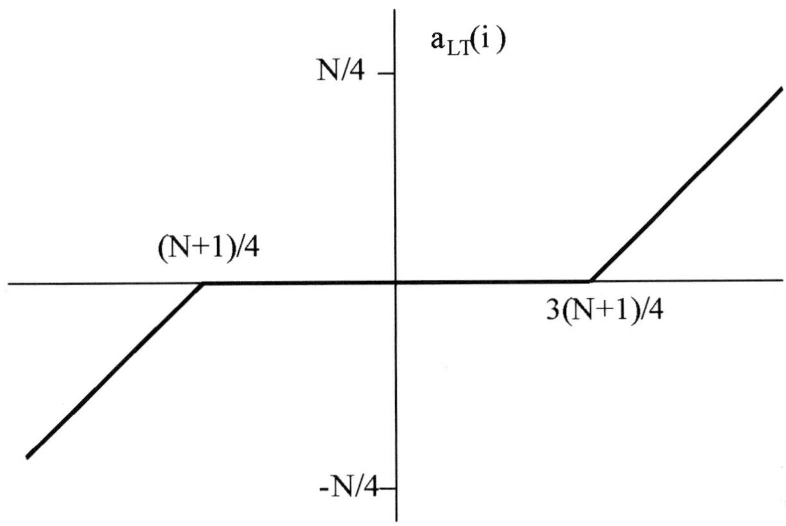

Figure 1.1. *Rank scores for a light-tailed distribution.*

These scores, which are shown in Figure 1.1, were among those suggested by Gastwirth (1965).

If the sample distribution appears to be skewed to the right, which is a common occurrence with many real-world data sets, we could use a test that is sensitive to location shifts with right-skewed data. One such set of scores is

$$a_{RS}(i) = \begin{cases} i - (N+1)/2 & \text{if } i < (N+1)/2, \\ 0 & \text{otherwise.} \end{cases}$$

These scores are shown in Figure 1.2. In their original formulation of the adaptive test, Hogg, Fisher, and Randles (1975) did not allow for distributions that are skewed to the left, but they did suggest modifications that would be appropriate if left-skewed distributions needed to be considered.

In the HFR test we compute the selection statistics Q_3 and Q_4 and then use them to determine the appropriate choice of rank scores. If $Q_3 \le 2$ and $2 \le Q_4 \le 7$, the Wilcoxon scores will be used to compute the test statistic $S = \sum_{j=1}^{n} a_W(R_j)$. If $Q_4 > 7$, the scores for the median test are used to compute $S = \sum_{j=1}^{n} a_M(R_j)$. If the distribution is not skewed and has light tails such that $Q_3 \le 2$ and $1 \le Q_4 < 2$, then the scores for light-tailed distributions will be used to compute $S = \sum_{j=1}^{n} a_{LT}(R_j)$. If the distribution appears to be so skewed that $Q_3 > 2$, and if $Q_4 \le 7$, then the scores for right-tailed distributions will be used to compute $S = \sum_{j=1}^{n} a_{RS}(R_j)$. These regions and their corresponding scores are indicated in Figure 1.3.

In order to perform the adaptive test we compute Q_3 and Q_4, select the appropriate rank scores, and compute the test statistic. We first compute the mean of the selected scores,

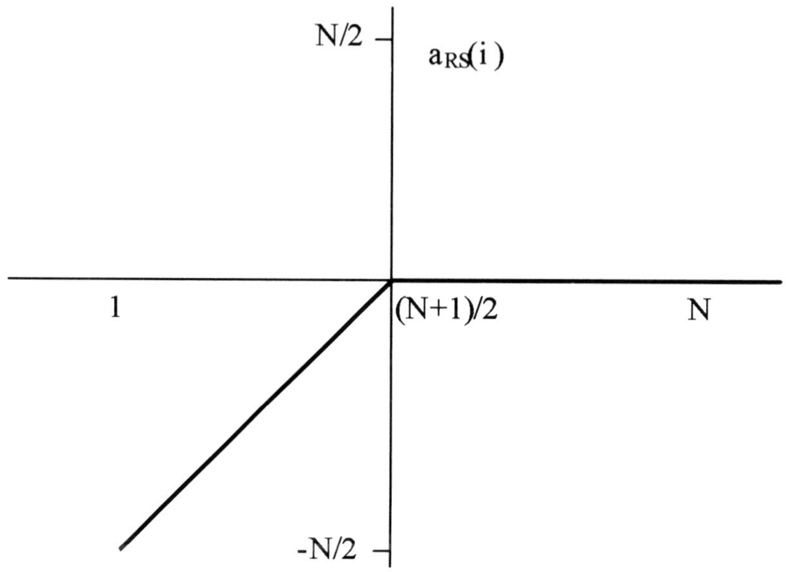

Figure 1.2. *Rank scores for a right-skewed distribution.*

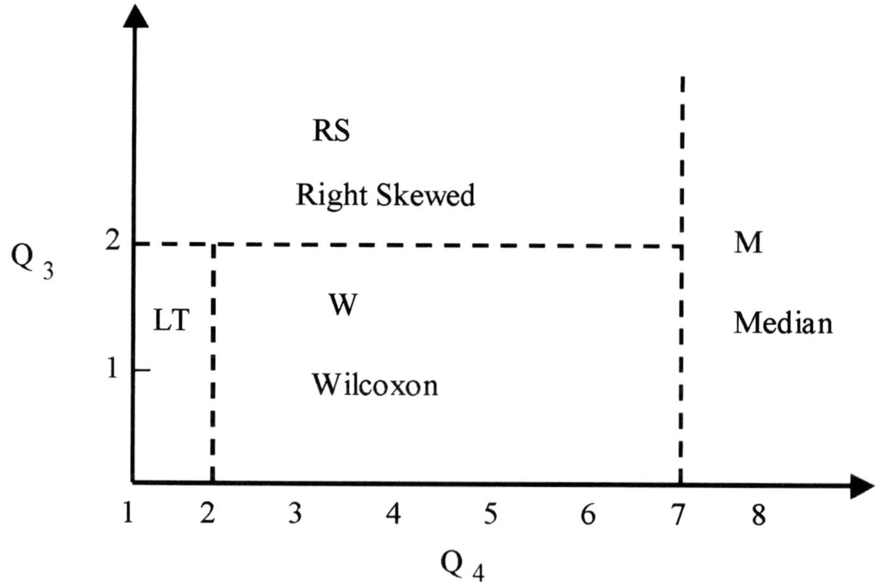

Figure 1.3. *Rank score regions for the HFR test.*

over all observations, as $\bar{a} = \sum_{i=1}^{N} a(i)/N$. It can be shown (see, for example, Randles and Wolfe (1979, Chap. 9)) that the standard error of S under the null hypothesis is

$$\sigma = \left\{ \frac{mn}{N(N-1)} \sum_{i=1}^{N} [a(i) - \bar{a}]^2 \right\}^{\frac{1}{2}}.$$

Next we compute the test statistic

$$Z = \frac{S - n\bar{a}}{\sigma}.$$

Since the test statistic has a limiting normal distribution, it is easy to obtain an approximate p-value. As we shall see, adaptive methods are generally not effective for small samples with $N < 10$, so the large sample approximation is usually sufficiently accurate for adaptive tests.

1.3.3 Why the HFR Test Maintains Its Significance Level

At first, it may appear to be "cheating" to use the data to determine which test should be used. However, it can be shown that the HFR test will have a Type I error that does not exceed the nominal value of α. In our discussion we let R indicate rejection of the null hypothesis and E_W be the event that the Wilcoxon scores are selected, which occurs if $Q_3 \le 2$ and $2 \le Q_4 \le 7$. Let E_M be the event that the median scores are selected, which occurs if $Q_4 > 7$. Let E_{LT} be the event that the scores for light-tailed distributions are selected, which occurs if $Q_3 \le 2$ and $1 \le Q_4 < 2$. Finally, let E_{RS} be the event that the scores for right-skewed distributions will be used, which occurs if $Q_3 > 2$ and $Q_4 \le 7$. Since the four regions in the (Q_3, Q_4) plane defined in the HFR test are mutually exclusive and exhaustive, we can write the probability of a Type I error as

$$\begin{aligned} P(R) &= P(R \cap E_W) + P(R \cap E_M) + P(R \cap E_{LT}) + P(R \cap E_{RS}) \\ &= P(R\,|E_W)P(E_W) + P(R\,|E_M)P(E_M) \\ &\quad + P(R\,|E_{LT})P(E_{LT}) + P(R\,|E_{RS})P(E_{RS}). \end{aligned}$$

Now consider $P(R\,|E_W)$, which is the probability of rejecting the null hypothesis when the Wilcoxon scores are used. The rejection region of the Wilcoxon test is determined by the distribution of the sum of the ranks in the second sample over all permutations of the ranks. Since, under the null hypothesis, the probability of observing any specified set of ranks equals the probability of observing any other set of ranks, and since it can be shown that Q_3 and Q_4 are independent of the ranks, the rejection region does not depend on Q_3 or Q_4. Hence, we can, in the adaptive test, use the same critical region for the Wilcoxon test as is commonly used in the nonadaptive Wilcoxon test. Because the distribution of S is discrete, we may not be able to obtain a Type I error rate exactly equal to α, but we can determine the rejection region, under the null hypothesis, so that $P(R\,|E_W) \le \alpha$.

We can make a similar argument for the other rank-based tests since their rejection regions also do not depend on Q_3 or Q_4. Consequently,

$$P(R) \le \alpha P(E_W) + \alpha P(E_M) + \alpha P(E_{LT}) + \alpha P(E_{RS}) = \alpha.$$

Hence, the adaptive test maintains the significance level α. In Chapter 8 we will give a more complete explanation of why rank-based tests maintain their level of significance. Note that this adaptive test maintained the significance level because each component of the test used rank-based test statistics and the rejection regions did not depend on Q_3 or Q_4. If the component tests were not based on ranks, the resulting test might not have maintained its significance level.

1.3.4 The Performance of the HFR Test

We have demonstrated that the actual significance level of the HFR test does not exceed α. Consequently, it is not surprising to find that its empirical significance level is close to $\alpha = .05$ for the simulation results shown in Table 1.1. These simulations used 10,000 data sets, with each data set having $m = n = 20$ observations. Nine distributions, each a member of the generalized lambda family of distributions, were used in these studies. The first three distributions are symmetric with a skewness of 0.0, the next three are moderately skewed with a skewness of 1.0, and the last are highly skewed with a skewness of 2.0. In this table we observe, as expected, that both the HFR test and the pooled two-sample t test maintain their significance level. The generalized lambda family of distributions, which was proposed by Ramberg et al. (1979), will be described in detail in the next chapter.

The corresponding empirical powers of the HFR test and the pooled two-sample t test are given in Table 1.2. In these simulation studies 1000 data sets were generated with $m = n = 20$ observations in each. For these power studies we set the difference (δ) between the two population means to 0.62 for each of the nine distributions used. This value of $\delta = 0.62$ was determined so that we would obtain approximately 50% power for

Table 1.1. *The empirical significance level in percent for the HFR test and the pooled t test with $m = n = 20$ observations, using a significance level of $\alpha = .05$.*

	Distributions								
Skewness	0	0	0	1	1	1	2	2	2
Kurtosis	1.8	3.0	6.0	4.2	5.4	8.4	11.4	12.6	15.6
HFR test	4.9	4.9	4.7	5.1	4.9	4.8	4.9	5.1	5.0
t test	4.9	5.0	4.9	4.8	4.9	5.0	4.7	4.6	4.6

Table 1.2. *The empirical power in percent for the HFR test and the pooled t test with $m = n = 20$ observations, using a significance level of $\alpha = .05$ with a shift of $\delta = 0.62$.*

	Distributions								
Skewness	0	0	0	1	1	1	2	2	2
Kurtosis	1.8	3.0	6.0	4.2	5.4	8.4	11.4	12.6	15.6
HFR test	51.0	49.0	56.2	57.3	55.3	58.7	74.6	72.5	66.8
t test	46.5	51.8	50.8	48.6	49.2	50.4	52.8	52.7	53.1

the pooled t test. The generalized lambda distribution with a skewness of zero and a kurtosis of 3.0 closely approximates a normal distribution. Note that the empirical power of the HFR test is slightly less than that of the pooled t test for the approximate normal distribution. This is not too surprising because the pooled t test is the most powerful test for normal populations with equal variances. However, note that the empirical power of the HFR test exceeds the power of the pooled t test for the other distributions and that the HFR test has a power advantage of 5% to 20% over the pooled t test for skewed distributions. Thus, for $m = n = 20$ it appears to be beneficial to use the adaptive test unless we know that the distribution is approximately normal.

1.4 Discussion of Rank-Based Adaptive Tests

Hogg, Fisher, and Randles (1975) published results for $m = n = 15$ for the HFR test that demonstrated the same properties as those shown in Tables 1.1 and 1.2. The results showed that the HFR test maintained its significance level and was often more powerful than traditional tests. Despite its good properties, the HFR test appears to be rarely used in practice. I believe that the failure of the HFR test to be more widely used is due to several factors.

One problem is that many researchers are naturally reluctant to use an adaptive method because it looks too much like cheating. It just doesn't seem right to use the data to select the test. I hope the reader will see that adaptive tests are valid if a great deal of care is taken in their construction to ensure that they really do maintain their significance level. This should encourage potential users to consider adaptive methods.

A more serious problem with the HFR adaptive test is that it could not easily be generalized to more complex models. Although it has been generalized to the one-way layout by Büning (1996), statistical consultants would like an adaptive procedure that could be used for two-factor designs, multiway designs, and multiple regression problems. The problem with the HFR test is that it is rank based, which makes it difficult to generalize. We can overcome this obstacle by using an adaptive method that uses a regression approach that can be applied to many situations. This regression approach will be described in Chapters 2 and 3.

Another problem with the HFR test is that there was no corresponding adaptive confidence interval. Adaptive confidence intervals require more calculations than adaptive tests, but they can be constructed in a manner that reduces the average width of the interval while maintaining the coverage probability when errors are nonnormal. An adaptive confidence interval procedure will be described in Chapter 6.

Also, the HFR test, while not difficult to perform, required some unusual calculations. The selection statistics Q_3 and Q_4 are tedious to compute by hand, and software was not available for calculating p-values for the HFR test. The adaptive methods described in this book are programmed in SAS macros listed in Appendix A and available on the World Wide Web at *www.math.niu.edu/~ogorman/*.

To sum up, the rank-based adaptive tests that have been available for many years are valid and effective, but their use is limited to some fairly simple tests of significance. More modern adaptive methods that are quite general will be fully described in this book. With these newer adaptive methods many of the difficulties associated with their use have been overcome. I believe that adaptive tests and confidence intervals are now practical alternatives to traditional statistical methods for a wide variety of applications.

1.5 The Relationship of Adaptive Methods to Robust Regression Methods

Adaptive methods and robust methods have similar objectives. Both methods attempt to increase efficiency over the OLS methods and both attempt to maintain reasonable efficiency when the usual regression assumptions are met. In addition, both adaptive and robust methods downweight the influence of outliers.

Although the adaptive methods and the robust methods have similar goals, the approaches used are quite different. Most of the robust methods have emphasized estimation of regression parameters in a linear regression model, while the adaptive methods have focused on tests of significance. Also, robust methods have tended to be nonadaptive in the sense that the influence function or trimming proportion is usually not adjusted, while the adaptive methods take an initial evaluation of the data before the appropriate technique is selected. The distinction between the two approaches is not always clear. For example, Dodge and Jurečková (2000) proposed an estimation procedure based on adaptively selecting from several robust methods.

There are several classes of robust regression estimators, including the M-estimators, the R-estimators, the least median of squares (LMS) estimators, and the least trimmed squares (LTS) estimators. Instead of minimizing the sum of the squared residuals, the M-estimators minimize the sum of some function of the residuals that is chosen to reduce the influence of outliers. For details on M-estimation see Hampel et al. (1986, Chap. 2). One problem with M-estimation is that there are several functions that could be chosen and each function has at least one tuning parameter that must be chosen. Once a function is selected the M-estimation procedure uses a fixed method to downweight outliers. Note that, if a large sample is taken, the M-estimation procedure will generally downweight some of the largest residuals, even if the error terms were normally distributed.

Another class of estimators are called R-estimators. These estimators are obtained by minimizing $\sum_{i=1}^{n} a\left(R_i\right) e_i$, where e_i is the ordinary residual, R_i is the rank of e_i, and $a(i)$ is a score function. For details of this approach see Rousseeuw and Leroy (1987, Chap. 3) or McKean and Sheather (1991). These M-estimators and R-estimators, which tend to downweight observations that are outliers, appear to be rarely used in practice. One problem with these robust estimators is that there are too many robust approaches, functions, and tuning parameters combined with too little guidance on their correct choice.

In the last few years the most popular robust regression methods appear to be based on the LMS and the LTS approaches. The LMS estimator minimizes the median of the squared residuals, whereas the OLS estimator minimizes the mean of the squared residuals. The LMS regression line, in simple linear regression, is the center of the narrowest strip that covers 50% of the data when the distances are measured vertically. This estimator is classified as a high breakdown point estimator because it is not greatly influenced by a large proportion of extreme outliers. However, as Ryan (1997, Chap. 11) has pointed out, the LMS estimator has low efficiency compared to traditional OLS methods when the usual regression assumptions are met.

Another high breakdown point estimator is the LTS estimator, which minimizes the sum of a proportion of the ordered squared residuals. The LTS estimator is similar to the OLS estimator except that some of the largest squared residuals are not used in the summation. This appears to be a promising method of robustifying regression estimates but, as Ryan (1997, Chap. 11) has indicated, the LTS estimators are computationally complex. It should also be noted that these methods are primarily estimation procedures; the finite sample properties of the corresponding tests and confidence intervals are rarely discussed in the literature.

Although robust and adaptive methods have similar objectives, we will concentrate on adaptive tests and adaptive confidence intervals. Robust methods have primarily addressed estimation problems, whereas adaptive methods have primarily addressed tests of significance. We will discuss robust estimation in greater detail in Chapter 7.

1.6 Informal Adjustments in the Analysis of Data

Some researchers prefer to adjust the data before they proceed with the analysis. They may use graphical methods to obtain a preliminary view of the data and then may choose to remove some outliers from the data before further analysis is performed. Another approach is to find a suitable transformation. Although this type of informal approach can often yield good results, it has several defects.

One problem is that, if a test of significance is to be performed, the informal adjustments may produce a test that fails to maintain its significance level. For example, if the researcher often discards observations that are set apart from the rest of the data, it is not clear that the resulting test will have a level of significance close to the nominal value. Furthermore, by removing some observations from the analysis, the testing procedure, when taken as a whole, may be viewed as a subjective process. The subjectivity of informal adaptation can be a serious problem because there may be a suspicion that the adjustments were made, consciously or unconsciously, in such a manner as to produce a desired result. However, if the analysis is being done for oneself, then the subjective nature of the informal procedure may not be a serious problem.

One goal of adaptive inference is to avoid the informal adjustments of data by using a predetermined method to downweight outliers. This is important because most serious statistical work is performed so that it can be published in scientific journals or used as part of a regulatory process. The advantage of adaptive methods is that the procedures maintain the objectivity that is necessary in scientific and regulatory work. The advantages and disadvantages of the traditional, adaptive, and informal approaches are summarized in Table 1.3.

1.7 Overview

This book contains an approach to adaptive testing and adaptive confidence intervals that can be used in many consulting situations. An adaptive two-sample test that is not based on ranks is introduced in Chapter 2 for the two-sample test. The adaptive weighting procedures used in this two-sample test are the same as those used in a general form of the adaptive test that is presented in Chapter 3. Several examples of adaptive tests are given in Chapter 4 along with

Table 1.3. *A comparison of some approaches to the analysis of data.*

Approach	Advantages	Disadvantages
Traditional	Robust for validity. Not subjective.	Not robust for efficiency with nonnormal errors.
Adaptive	Robust for validity. Robust for efficiency. Not subjective.	Requires more computation than traditional approaches.
Informal adjustments	Potential of improving the analysis.	Significance level unknown. Subjective.

detailed instructions on the use of the statistical software needed to compute them, using the SAS macros contained in Appendix A. In Chapter 5 the adaptive test for paired data is explained, which requires some modification of the general testing procedure. We next consider techniques for constructing adaptive confidence intervals, which are described in Chapter 6. These adaptive confidence intervals use the same weighting procedure as was used in the tests described in Chapters 2 and 3. In Chapter 7 we review some robust methods of estimation and present an adaptive estimation procedure for estimating any parameter in a linear model. Chapter 8 contains several adaptive methods including an adaptive multiple comparison procedure for one-way layouts, which can be performed using the general adaptive testing method. In Chapter 8 we also describe Büning's (1996) rank-based test for the one-way layout and address the robustness of the adaptive method when ties are present or when unequal variances are found in the populations.

A skeptic may ask, "Is it really possible to construct an adaptive test that maintains its significance level and is more powerful than the traditional F test?" To demonstrate to the reader that improved efficiency is possible with adaptive methods, I have included results from many simulation studies in Chapters 2, 4, and 5. These simulation studies were designed to estimate the significance level and power of the adaptive tests. Also, I have included coverage probabilities and average widths in Chapter 6 in order to demonstrate that confidence intervals can be made narrower while still maintaining their coverage probabilities. I hope this material will convince the reader that adaptive methods are often more efficient than traditional methods.

Chapter 2

An Adaptive Two-Sample Test

2.1 Two-Sample Tests

In Chapter 1 we described the HFR test, which was an attempt to improve the power of the two-sample test by using an adaptive method of selecting a set of rank scores. The major problem with the rank-based approach is that rank-based tests cannot easily be generalized to be used with more complex designs. For example, while it is easy to generalize a two-sample rank-based test to a one-way layout, it is difficult to generalize the rank-based test to test the interaction effect in a two-way layout.

Instead of using ranks we will use a weighting procedure to improve the performance of the methods. We begin our discussion with an example that concerns psychological factors related to the progression of cancer. This data set, which was originally given by Blumberg, West, and Ellis (1954), was analyzed in Lehmann (1975) using nonparametric methods. In this data set one group had a rapidly progressing form of the disease while the other group had a slowly progressing form. Each subject was given a test that produced a score. Highly negative scores were indicative of a tendency to present the appearance of serenity in the presence of stress. The scores for the two groups are given in Table 2.1.

A dot plot of these observations, given in Figure 2.1, shows that these data sets are skewed to the right with the scores in the second group tending to be larger than those in the first group. Let n equal the total number of observations. The analysis will use a linear model for data from two populations, which can be formulated as

$$y_i = \beta_0 + \beta_1 x_i + \varepsilon_i$$

for $i = 1, \ldots, n$, where $y_i =$ the score for the ith observation and

$$x_i = \begin{cases} 1 & \text{if the } i\text{th subject is from group 2,} \\ 0 & \text{if the } i\text{th subject is from group 1.} \end{cases}$$

With this parameterization, the parameter β_0 is the mean of population 1 and $\beta_0 + \beta_1$ is the mean of population 2. A two-tailed test of $H_0 : \beta_1 = 0$ versus $H_a : \beta_1 \neq 0$ can be performed by using a test statistic for a pooled two-sample t test that can be obtained in several ways. One way is to use the linear model and compute the usual test statistics for

Table 2.1. *Cancer data set.*

Group 1	−25, −24, −22, −22, −21, −18, −18, −18, −18, −17, −16, −14, −14, −13, −13, −13, −13, −9, −8, −7, −5, 1, 3, 7, 7
Group 2	−21, −18, −16, −16, −16, −14, −13, −13, −12, −11, −11, −11, −9, −9, −9, −9, −7, −6, −3, −2, 3, 10

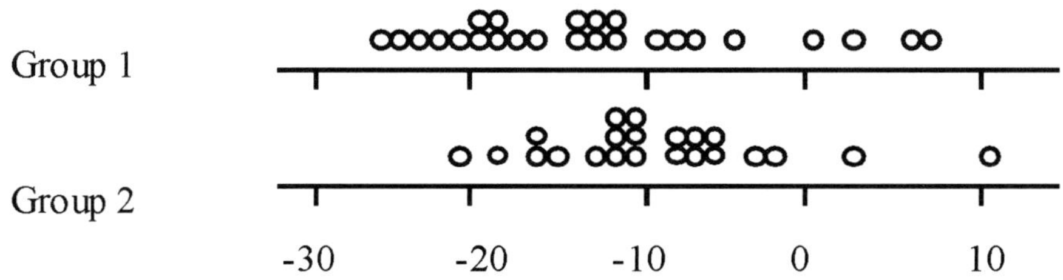

Figure 2.1. *A dot plot of the cancer data.*

testing $H_0 : \beta_1 = 0$ versus $H_a : \beta_1 \neq 0$. The same test statistic can be obtained by using the formula for the pooled t test statistic. The advantage of the linear model approach is that it can be generalized to more complex models.

If the errors are normally distributed, are independent, and have constant variance, then the t test is the uniformly most powerful test within the class of all unbiased tests. However, because we believe that the errors may not be normally distributed, we may be interested in finding a test with greater power than the t test for nonnormal error distributions. Our adaptive approach will be to observe how closely our data fit a normal error model. If the data are consistent with the normal error model, then we do not want to greatly modify the traditional procedure. Conversely, if the data appear to depart considerably from the normal error model, we will attempt to improve the power of the test by using an appropriate weighting of the observations.

2.2 Assessing Normality

We begin our adaptive procedure by observing how closely our data conform to the normal distribution. Since we will perform this check on the data under the null hypothesis, which is the statement that the population means are equal, we can combine the data from the two groups to assess normality. In our adaptive test we will use the deleted residuals, computed under the null hypothesis, to determine how much weight should be given to each observation.

The deleted residual for the ith observation, which will be denoted by $e_{(i)}$, is computed by removing the ith observation from the data set before computing the predicted value for the ith observation. Thus, $e_{(i)} = y_i - \hat{y}_{(i)}$, where $\hat{y}_{(i)}$ is the predicted value based on the $n - 1$ remaining observations. The *studentized deleted residual* for the ith observation, which we will denote by d_i, equals the deleted residual divided by the estimated standard deviation of the deleted residual (see Neter, Wasserman, and Kutner (1989, Chap. 11)). We will express d_i in a general form in the next chapter, but for the two-sample test we can calculate the studentized deleted residual as

$$d_i = \frac{e_{(i)}}{s\{e_{(i)}\}} = \frac{e_{(i)}}{\sqrt{[n/(n-1)]\, MSE_{(i)}}},$$

where the mean squared error $MSE_{(i)}$ is based on the $n-1$ observations remaining after removing the ith observation. In the next chapter we will give a shortcut formula for the calculation of d_i that does not require the recalculation of $\hat{y}_{(i)}$ and $MSE_{(i)}$ for each observation.

For example, if we want to compute the studentized deleted residual for $i = 2$ in our cancer data set with $n = 47$ observations, we first compute $\hat{y}_{(2)}$ based on the model $y_i = \beta_0 + \varepsilon_i$ for $i = 1, 3, 4, \ldots, 47$. Since d_i is computed under the null hypothesis, the predicted value $\hat{y}_{(2)} = -10.848$ is simply the average of the observations that remain after y_2 is removed. Consequently, the deleted residual is $e_{(2)} = y_2 - \hat{y}_{(2)} = -24 - (-10.848) = -13.152$. We then compue $s\{e_{(2)}\}$ using

$$MSE_{(2)} = \frac{1}{n-2} \sum_{i=1,\, i\neq 2}^{n} \left(y_i - \hat{y}_{(2)}\right)^2 = \frac{1}{45}\,(2961.93) = 65.821.$$

Hence,

$$d_2 = \frac{e_{(2)}}{\sqrt{[n/(n-1)]\, MSE_{(2)}}} = \frac{-13.152}{\sqrt{(47/46)\, 65.821}} = \frac{-13.152}{8.201} = -1.604.$$

The residuals and the studentized residuals are given in Table 2.2 for some of the observations in the data set. These were computed using PROC GLM in SAS with a model that only included the intercept.

One reason we use the studentized deleted residuals is that, if the errors are normally distributed, each d_i will have a t distribution with $n-2$ degrees of freedom (ν). Although the studentized deleted residuals are not independent, we can get some idea of how closely the errors follow the normal distribution by comparing the studentized deleted residuals to t variates with $\nu = n-2$.

2.3 Smoothing Residuals

The next step is to determine how far these studentized deleted residuals depart from what we would expect if the errors were normally distributed. The empirical c.d.f. of the studentized deleted residuals from the cancer data set is shown in Figure 2.2. Note that the empirical c.d.f. is uneven in appearance, which makes it difficult to compare it to a t distribution. In order to get a better idea of its shape, we will smooth the empirical c.d.f. of the studentized deleted residuals by using a smoothing function. The amount of smoothing is determined by the bandwidth. We will use a bandwidth of $h = 1.587\,\hat{\sigma}\, n^{-1/3}$, which has the property that it will minimize the mean integrated squared error when the observations are independent

Table 2.2. *The residuals and the studentized deleted residuals for some of the observations in the cancer data set.*

Observation	Scores	Group	Residual	Stud. Del. Res.
1	−25	1	−13.8723	−1.73640
2	−24	1	−12.8723	−1.60378
3	−22	1	−10.8723	−1.34364
4	−22	1	−10.8723	−1.34364
.	.	.	.	.
.	.	.	.	.
.	.	.	.	.
23	3	1	14.1277	1.77057
24	7	1	18.1277	2.32482
25	7	1	18.1277	2.32482
26	−21	2	−9.8723	−1.21578
27	−18	2	−6.8723	−0.83926
28	−16	2	−4.8723	−0.59271
.	.	.	.	.
.	.	.	.	.
.	.	.	.	.
44	−3	2	8.1277	0.99567
45	−2	2	9.1277	1.12141
46	3	2	14.1277	1.77057
47	10	2	21.1277	2.76983

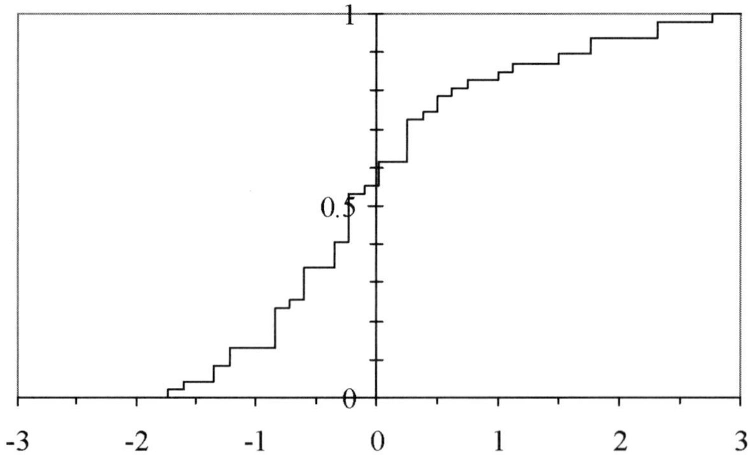

Figure 2.2. *The empirical c.d.f. of the studentized deleted residuals for the cancer data set.*

normals (see Polansky (1998)). Since the observations are studentized, the variance should not depart too much from $\sigma^2 = 1$, so a bandwidth of $h = 1.587\ n^{-1/3}$ will be used to obtain the smoothed distribution function. The smoothed c.d.f. at point d over the set of all studentized deleted residuals (D) is computed as

$$\hat{F}_h(d; D) = \frac{1}{n} \sum_{i=1}^{n} \Phi \left(\frac{d - d_i}{h} \right),$$

where $\Phi(\cdot)$ is the c.d.f. of the standard normal distribution.

In order to adjust the studentized deleted residuals so that they correspond to the t distribution, we will center them by subtracting the estimated median $(\tilde{d})$ from each studentized deleted residual. The value of $\tilde{d}$ is determined by a search process so that $\hat{F}_h(\tilde{d}; D) = .5$, and the *centered studentized deleted residuals* are calculated as $d_{c,i} = d_i - \tilde{d}$. For brevity, the centered studentized deleted residuals will be called the *residuals* and the set of these residuals will be indicated by $D_c = \{d_{c,1}, \ldots, d_{c,n}\}$. The smoothed c.d.f. of the residuals, which will be denoted as $\hat{F}_h(d; D_c)$, is plotted in Figure 2.3 for the cancer data set along with the c.d.f. of the t distribution.

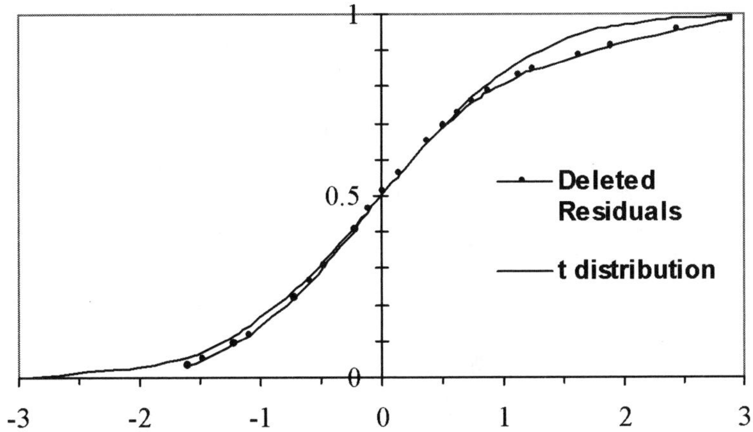

Figure 2.3. *The smoothed c.d.f. of the centered studentized deleted residuals for the cancer data set and the c.d.f. of the t distribution with $v = n - 2$ degrees of freedom.*

We note that, for the cancer data set, the residuals with $d_{c,i} > 1$ tend to be larger than what we would expect if the error distribution had been normal, and the residuals with $d_{c,i} < -1$ tend to be slightly smaller than what we would expect had the error distribution been normal. We would like to transform the data so that the c.d.f. of the transformed residuals is closer to the t distribution. This will have the effect of making the errors more closely follow the normal distribution. To accomplish this objective we will weight each observation depending on the residual and the corresponding percentile of the t distribution.

2.4 Weighting Observations

The weighting procedure will be illustrated by considering the 25th observation in the data set, which, when used in the model, produced the second-largest residual. By fitting the model under the null hypothesis we computed the studentized deleted residuals as $d_{25} = 2.325$ (see Table 2.2). An iterative procedure was then used to determine the median $\tilde{d} = -0.131$, which was then used to compute the centered studentized deleted residual. Thus, after centering, we found $d_{c,25} = 2.325 - (-0.131) = 2.456$. We then computed the smoothed c.d.f. for this residual to be $\hat{F}_h\left(d_{c,25}; D_c\right) = .95558$ by using the smoothing procedure with a bandwidth of $h = 1.587n^{-1/3} = 1.587\,(47)^{-1/3} = 0.440$. In the weighting procedure we will compare the c.d.f. of the residuals to the c.d.f. of the t distribution with $df = n - 2$, which will be denoted by $T_{n-2}(\cdot)$. Let t_i denote the t variate for the ith observation such that $T_{n-2}(t_i) = \hat{F}_h\left(d_{c,i}; D_c\right)$. So for the 25th observation we determined the t variate such that $T_{n-2}(t_{25}) = \hat{F}_h\left(d_{c,25}; D_c\right) = .95558$. By using an algorithm to compute the inverse t distribution we obtained $t_{25} = T_{45}^{-1}[\hat{F}_h(d_{c,25}; D_c)] = 1.739$.

Consequently, had the errors been normal, we would have expected to obtain a residual near $t_{25} = 1.739$ for the second-largest residual with $n = 47$ observations. Since our objective is to weight the observations so that the weighted residual is close to 1.739 and since we observed a residual of $d_{c,25} = 2.456$, we weight this observation by

$$w_{25} = \frac{t_{25}}{d_{c,25}} = \frac{1.739}{2.456} = 0.708.$$

This weight will be used for the 25th observation in a WLS regression model. We are downweighting the observation because the residual $d_{c,25}$ is much larger than t_{25}. If t_i approximated $d_{c,i}$, then the weight would be near one.

For the second observation, $y_2 = -24$, we computed the deleted studentized residual of $d_2 = -1.604$. After centering with $\tilde{d} = -0.131$ we found $d_{c,2} = -1.604 - (-0.131) = -1.473$, which is the second-smallest residual plotted in Figure 2.3. The smoothed c.d.f. is $\hat{F}_h\left(d_{c,2}; D_c\right) = .04963$ so that $t_2 = T_{45}^{-1}\,[0.0493] = -1.683$. Thus, the weight is

$$w_2 = \frac{t_2}{d_{c,2}} = \frac{-1.683}{-1.473} = 1.142.$$

We will increase the weight of this observation because $d_{c,2}$ is closer to zero than t_2.

Using the notation that we have introduced, we can express the weights for the adaptive two-sample test as

$$w_i = \frac{t_i}{d_{c,i}} = \frac{T_{n-2}^{-1}(\hat{F}_h(d_{c,i}; D_c))}{d_{c,i}}$$

for $i = 1, \ldots, n$. If the c.d.f. of the residuals approximates $T_{n-2}(\cdot)$, then all of the observations will be given weights near one so that the resulting test will be similar to the traditional test. However, if the c.d.f. of the studentized deleted residuals departs greatly from the c.d.f. of the t distribution, then the observations will be weighted so that, after weighting, the distribution of the studentized deleted residuals will more closely approximate the t distribution. By referring to Figure 2.3 we can see, for our example using the cancer data set, that many of the larger residuals need to be reduced in size so that their c.d.f. will approximate

$T_{n-2}(\cdot)$, and the smaller residuals need to be increased in size. This is accomplished by giving weights less than one to the observations with the largest residuals and greater than one to observations with the most highly negative residuals.

The rationale for using the weights $w_i = t_i / d_{c,i}$ can be illustrated with an example. Suppose we have a large data set and we compute the smoothed c.d.f. of the centered studentized deleted residuals. Further, suppose we find that the smoothed c.d.f. of these residuals equals the c.d.f. of the t distribution for all observations except the last observation and, for the last observation, the centered studentized residual is twice as large as the corresponding t value. That is, we have $d_i = t_i$ for $i = 1, \ldots, n-1$ and $d_n = 2t_n$ so that the weights are $w_i = 1$ for $i = 1, \ldots, n-1$ and $w_n = 1/2$. Because the nth observation is an outlier it will make a larger contribution to $SSE = \sum_{i=1}^{n} (y_i - \hat{y}_i)^2$ than it would have had the empirical c.d.f. agreed with the t distribution for all observations. To reduce the influence of the last term $(y_n - \hat{y}_n)^2$ in the sum of squared errors we multiply it by the weight w_n so that the weighted sum of squared errors is

$$SSE^* = (y_1 - \hat{y}_1)^2 + \cdots + (y_{n-1} - \hat{y}_{n-1})^2 + \left[\frac{1}{2}(y_n - \hat{y}_n)\right]^2.$$

After weighting the observations, the terms of SSE^* now approximate what the terms of the sum of squared errors would have been had the errors been normally distributed. It is easy to minimize SSE^* since the weighted sum of squared errors is minimized when the WLS method is used.

In actual practice with real data sets the weights will deviate from one, which complicates the analysis of the effect of the weighting. Also, we have used the deleted studentized residuals rather than the ordinary residuals in our formulation of the weights. In addition, the weights themselves do not need to be correct, they need only reflect a reasonable weighting relative to other observations in the data set. Multiplying all the weights by a constant will not change any of the test statistics or estimates. Other weighting procedures may be used, but the proposed weights produce tests with good performance relative to the t test.

After we have computed the weights w_i for $i = 1, \ldots, n$, they can be used in a WLS regression analysis. Using our model for the two-sample data $y_i = \beta_0 + \beta_1 x_i + \varepsilon_i$, we obtain the WLS estimates of β_0 and β_1 by minimizing the weighted sum of squared errors

$$\sum_{i=1}^{n} \{w_i [y_i - (\beta_0 + \beta_1 x_i)]\}^2 = \sum_{i=1}^{n} \{w_i y_i - (\beta_0 w_i + \beta_1 w_i x_i)\}^2.$$

This weighted sum of squared errors will be minimized at the value SSE_c^* for the complete model. One way to find SSE_c^* is to use the transformed variables $y_i^* = w_i y_i$ and $x_i^* = w_i x_i$ with the OLS procedure to minimize

$$\sum_{i=1}^{n} \left\{y_i^* - \left(\beta_0 w_i + \beta_1 x_i^*\right)\right\}^2.$$

If these transformations are effective, the transformed model should have approximately normal errors. To investigate how effective this transformation was with the cancer data set, we computed the studentized deleted residuals for the transformed model. Figure 2.4 gives

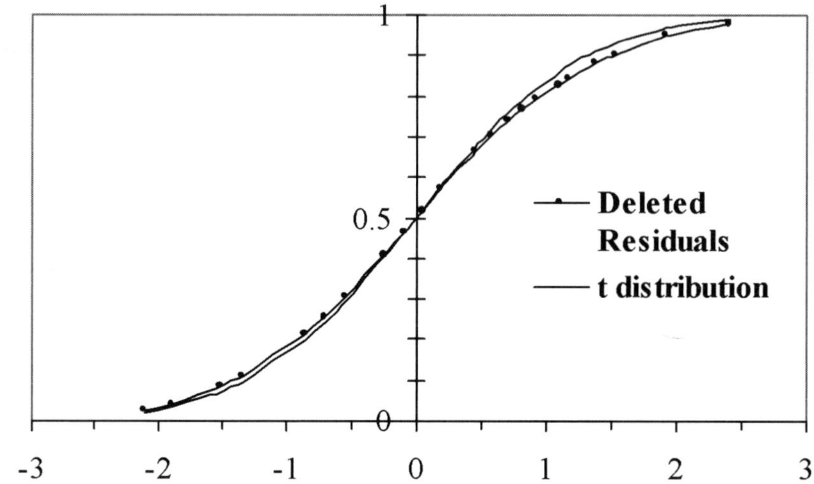

Figure 2.4. *The smoothed and centered c.d.f. of the studentized deleted residuals for the cancer data set after weighting and the c.d.f. of the t distribution with* $\nu = n - 2$ *degrees of freedom.*

the smoothed and centered c.d.f. of the studentized deleted residuals after the observations were weighted. Note that this c.d.f. is reasonably close to the c.d.f. of the t distribution with $\nu = n - 2 = 45$.

We now need to calculate a test statistic to test $H_0 : \beta_1 = 0$ versus $H_a : \beta_1 \neq 0$. Under the null hypothesis $H_0 : \beta_1 = 0$ the model becomes

$$y_i = \beta_0 + \varepsilon_i .$$

This model, which was used to calculate the weights, will be called the reduced model. We can use WLS to minimize the weighted sum of squared errors for the reduced model as

$$\sum_{i=1}^{n} \{ w_i [y_i - \beta_0] \}^2 = \sum_{i=1}^{n} \left(y_i^* - \beta_0 w_i \right)^2 .$$

This minimum value will be indicated by SSE_R^*. To test $H_0 : \beta_1 = 0$ against $H_a : \beta_1 \neq 0$, we will use SSE_C^* from the complete model to compute the test statistic

$$F^* = \frac{(SSE_R^* - SSE_c^*)/1}{SSE_C^*/(n-2)} .$$

This is the same test statistic as is usually used to test subsets of regression coefficients in linear models, but it will not be used in the usual manner. If SSE_R^* is much greater than SSE_C^*, it is an indication that the reduced model $y_i = \beta_0 + \varepsilon_i$ does not provide a good fit compared to the complete model $y_i = \beta_0 + \beta_1 x_i + \varepsilon_i$. Thus, we will reject the null hypothesis for large values of F^*. The next step in the testing process is to determine how large F^* needs to be before we reject the null hypothesis. This is not easy to determine because the F statistic may not have an F distribution, even if the usual inference assumptions are satisfied, due to the adaptive nature of the test. In order to construct a fair test that maintains its significance level we need to use a permutation technique.

2.5 Performing a Permutation Test

Permutation tests require the user to permute, or shuffle, the observations in order to determine a p-value. These tests have a long history, dating back to an article by Pitman (1937), who stated that the permutation approach was implicit in Fisher's earlier writings. For a two-sample test, the full permutation test requires that the test statistic be computed for every permutation of the assignment of the units to the two groups, such that the number of observations in each group is unchanged. If the number of permutations is large, sample permutation methods can be used to estimate a p-value, but even the sample permutation tests were not practical until fast computers and good software were commonly available. Consequently, permutation tests were rarely used in the half century following the publication of Pitman's work in 1937. Nowadays, the situation is quite different, so we can often use computationally intensive methods without too much worry about time or cost. Most of the adaptive tests in this book use sample permutation methods that can be performed in a few minutes.

We first note that, under the null hypothesis, the scores are not related to group membership. Thus, under the null hypothesis, any permutation (or shuffling) of the scores is as likely as any other. When we used the complete model $y_i = \beta_0 + \beta_1 x_i + \varepsilon_i$, we coded it for the original data as follows:

$$\begin{aligned} y_1 &= \beta_0 && + \varepsilon_1, \\ &\ \vdots \\ y_{25} &= \beta_0 && + \varepsilon_{25}, \\ y_{26} &= \beta_0 + \beta_1 x_{26} && + \varepsilon_{26}, \\ &\ \vdots \\ y_{47} &= \beta_0 + \beta_1 x_{47} && + \varepsilon_{47}. \end{aligned}$$

When we randomly assign 25 individuals to group 1 we are, in effect, using the model $y_i = \beta_0 + \varepsilon_i$ for those individuals, and we are using the model $y_i = \beta_0 + \beta_1 + \varepsilon_i$ for the remaining 22 individuals. Thus, we can view a permutation as either a shuffling of the scores or a shuffling of the indicator variables. It will be more convenient, in the general adaptive test procedure that will be described in Chapter 3, to shuffle the indicator variables.

We recall that when we compute the weights under the null hypothesis $H_0 : \beta_1 = 0$ we use the reduced model $y = \beta_0 + \varepsilon_i$, which does not change if we shuffle the indicator variable. Consequently, after we randomly assign individuals to groups, we use the weights computed under the null hypothesis to transform the model and then compute the sum of squared errors for the full model. Let $SSE^*_{C,i}$ be the sum of squared errors for the ith permutation. Since SSE^*_R remains unchanged with the permutations of group membership, we can easily compute, for the ith permutation, the test statistic

$$F^*_i = \frac{SSE^*_R - SSE^*_{C,i}}{SSE^*_{C,i}/(n-2)}.$$

In order to exactly determine if the observed F^* is so large that we should reject the null hypothesis, we need to compute the values of F_i^* for all permutations. For the cancer data set we would need to compute F_i^* for $\binom{47}{25} = 1.48 \times 10^{13}$ permutations. Even if we could compute F_i^* for 100,000 permutations per second, it would take over four years to compute the full permutation distribution! Clearly, another approach is needed.

Instead of computing the full permutation distribution of the test statistic, we will take a large number (R) of random permutations to obtain a reasonably accurate estimate. We will use $R = 2000$ random permutations for most simulations in this book, but larger values of R can be used. These 2000 permutations will give us a reasonably accurate idea of how large F^* must be before we reject the null hypothesis. We can compute an approximate p-value from our sample of 2000 permutations by counting the number of permutation test statistics that exceed the observed test statistic. If F^* is so large that few permutation test statistics exceed F^*, then we have good reason to reject the null hypothesis. Let E equal the number of permutations with test statistics that equal or exceed the observed test statistic F^*. For this sample permutation test we can compute an approximate p-value with $p = (E + 1)/(R + 1)$, as recommended by Davison and Hinkley (1997). If the p-value is less than or equal to the significance level of α, we will reject the null hypothesis.

Since we used the reduced model to compute the weights, the $SSE_{c,i}^*$ are equally likely under the null hypothesis and are as likely to be observed as SSE_c^*. Hence, the formula $p = (E + 1)/(R + 1)$ can be justified by noting that the observed test statistic F^* is computed from one of the $R + 1$ equally likely assignments of the 47 individuals to the groups. If we had used the complete model to obtain the weights, we could not have used this formula because the observed test statistic would have been computed from the original assignment that produced the weights, while the other permutation test statistics would have been computed from permutations that were not used to calculate the weights.

2.6 Discussion of the Example

The adaptive test procedures can be illustrated with the data on the psychological factors related to cancer. To compute the p-values we computed F^* from the unpermuted data and, by counting the $R = 2000$ permutations with $F_i \geq F^*$, we found $E = 136$. Consequently, the p-value is $p = (136+1)/(2000+1) = .068$ for this data set. This p-value is considerably smaller than the p-value of .264 based on the pooled t test from the unweighted regression. While it is interesting to note the difference in p-values between the traditional tests and the adaptive test, it is not possible to conclude that the adaptive method is more powerful because we do not really know if there is a difference between these two groups. We now turn to simulation studies to evaluate the effectiveness of the adaptive test.

2.7 The Performance of the Adaptive Two-Sample Test

2.7.1 Description of the Simulation Study

Monte Carlo methods were used to evaluate the performance of the proposed test. For each distribution and sample size studied there were two simulations: one designed to estimate the size of the test and the other to estimate the power of the test. For most of the simulations we used a significance level of $\alpha = .05$ and used 10,000 data sets to check the size and to estimate the power of the tests. For those simulations that used 10,000 data sets with $\alpha = .05$, the standard error of the percentage of rejections under the null hypothesis was 0.22%. For each data set generated, 2000 permutations were used to calculate the p-value for the adaptive test. In addition, we used 100,000 data sets for a few special situations in order to get a more accurate estimate of the sizes of the adaptive and traditional tests.

The distributions used to generate the observations for these simulations were members of the generalized lambda family of distributions (Ramberg et al. (1979)), which has a four-parameter inverse c.d.f. of

$$F^{-1}(u) = \lambda_1 + \left[u^{\lambda_3} - (1-u)^{\lambda_4}\right]/\lambda_2.$$

The location parameter is λ_1 and the scale parameter is λ_2. The parameters λ_3 and λ_4 are complicated functions of the mean, standard deviation, skewness, and kurtosis. In this book skewness (α_3) is defined as $E(x-\mu)^3/\sigma^3$ and kurtosis (α_4) is defined as $E(x-\mu)^4/\sigma^4$. For each distribution used in the simulation we used the formulas in Ramberg et al. (1979) to compute the parameters $(\lambda_1, \lambda_2, \lambda_3, \lambda_4)$ based on specified values of the mean, standard deviation, α_3, and α_4.

The generalized lambda distributions were used because the skewness and kurtosis could be specified. For example, if we want to obtain a distribution whose first few moments equal those of the standard normal distribution, we set $\alpha_1 = 0, \alpha_2 = 1, \alpha_3 = 0$, and $\alpha_4 = 3.0$. For this distribution we find the parameters $\lambda_1 = 0$, $\lambda_2 = 0.1974$, $\lambda_3 = \lambda_4 = 0.1349$, which are then used in the inverse c.d.f. of the generalized lambda distribution. This approximate normal distribution is plotted in Figure 2.5. The similarity between this distribution and the standard normal is not too surprising because the first four moments match those of the standard normal. In addition, Ramberg et al. (1979) stated that the maximum absolute difference between the cumulative distribution functions of these distributions is approximately 0.001, so it seems reasonable to use this approximate normal distribution to generate values from symmetric distributions with moderate tails.

In order to generate variates from a long-tailed symmetric distribution, we set $\alpha_1 = 0$, $\alpha_2 = 1$, $\alpha_3 = 0$, and $\alpha_4 = 6.0$, which are the first four moments of the double-exponential

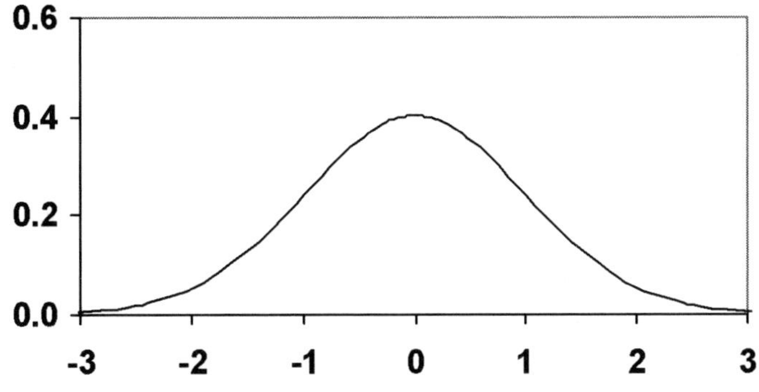

Figure 2.5. *Density function for the approximate normal distribution with $\alpha_3 = 0$ and $\alpha_4 = 3.0$.*

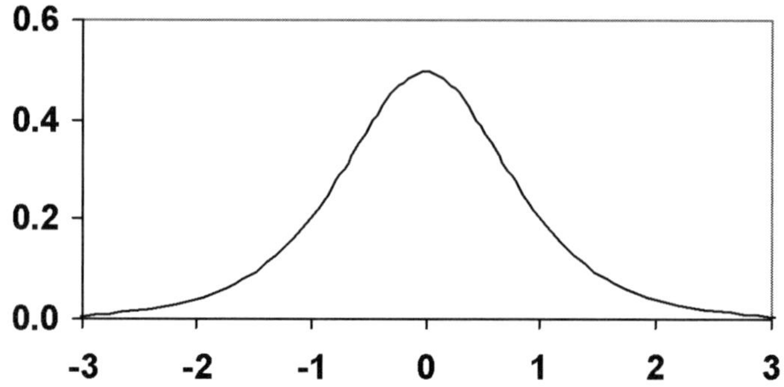

Figure 2.6. *Density function for the long-tailed symmetric distribution with* $\alpha_3 = 0$ *and* $\alpha_4 = 6.0$.

distribution. Again, we use the formulas in Ramberg et al. (1979) to obtain λ_1, λ_2, λ_3, and λ_4, which when used in the inverse c.d.f. of the generalized lambda distribution produce the distribution shown in Figure 2.6. This long-tailed distribution was used to generate data for sets that contain outliers.

The generalized lambda distribution can also be used to generate variates from highly skewed distributions. If we set $\alpha_1 = 0$, $\alpha_2 = 1$, $\alpha_3 = 2.0$, and $\alpha_4 = 12.6$, we obtain distributional parameters for the skewed distribution shown in Figure 2.7. This is an important distribution to use in the simulation study because many real-world data sets are skewed.

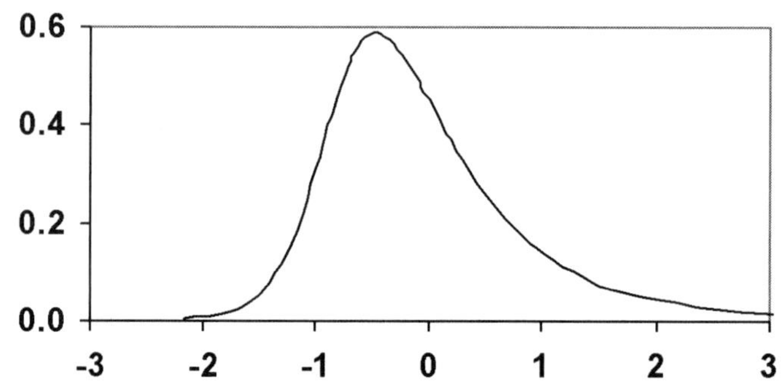

Figure 2.7. *Density function for the highly skewed distribution with moderate kurtosis with* $\alpha_3 = 2$ *and* $\alpha_4 = 12.6$.

Nine distributions were used in this simulation study, with each distribution having a mean of zero and a standard deviation of one so that $\alpha_1 = 0$ and $\alpha_2 = 1$. For the three symmetric distributions with $\alpha_3 = 0.0$ we obtain the uniform distribution when we use $\alpha_4 = 1.8$, the approximate normal distribution when we use $\alpha_4 = 3.0$, and a long-tailed distribution when we use $\alpha_4 = 6.0$. In order to generate variates from the three moderately skewed distributions, we set $\alpha_3 = 1.0$. For the low kurtosis distribution we use $\alpha_4 = 4.2$, for the moderate kurtosis distribution we use $\alpha_4 = 5.4$, and for the high kurtosis distribution we use $\alpha_4 = 8.4$. To generate variates from the three highly skewed distributions we set

$\alpha_3 = 2.0$. For the low kurtosis distribution we set $\alpha_4 = 11.4$, for the moderate kurtosis distribution we set $\alpha_4 = 12.6$, and for the high kurtosis distribution we set $\alpha_4 = 15.6$. These nine distributions, which represent a variety of symmetric and skewed distributions, are used throughout this book to generate error terms for various simulation studies.

2.7.2 The Size of the Adaptive Test

We need to check the empirical size of the adaptive test and the t test because, for most of the situations we will study, there is no proof that either of these tests will maintain their size. The pooled t test will maintain its size for models with normal errors, but in practice we rarely know that the errors are normally distributed. Nonadaptive permutation tests will maintain their size, but since the observations are adaptively weighted in the adaptive test, it is not clear that it will maintain its size. Hence, we will be looking carefully at the empirical size of both tests.

For the two-sample tests let n_1 be the number of observations in the first sample and let n_2 be the number of observations in the second sample. The equal sample size configurations that were studied in the simulations were $(n_1, n_2) = (6, 6), (10, 10), (20, 20)$, and $(50, 50)$. The unequal sample size configurations were $(n_1, n_2) = (6, 25)$ and $(48, 12)$.

For each sample size configuration, simulation studies were performed for each of the nine generalized lambda distributions used to generate the observations. The error distributions used in the two-sample test, which have a mean of zero and a variance of one, are described by their skewness and kurtosis in the column headings of Table 2.3. Note that the first three of these are symmetric, with the second distribution closely approximating the standard normal distribution. In order to estimate the size, the difference between the two groups was set equal to zero for the size simulations. The empirical sizes for these simulations are shown in Table 2.3.

An examination of the empirical sizes of the adaptive test and the pooled t test shown in Table 2.3 indicates that both tests maintain their size very close to the nominal value of 5%. Because the standard error of these estimates equals 0.22%, we would expect that approximately 95% of these estimates would be in the range $5\% \pm 2(0.22)\% = [4.56\%, 5.44\%]$. In only one of the 54 simulations did the empirical size of the adaptive test fall outside that range, and it was 5.6%. However, in 14 of the 54 simulations the empirical size of the pooled t test fell outside that range, indicating that the pooled t test does not always maintain its size when used with data generated from nonnormal distributions. Nevertheless, since the empirical size of the t test was never far from the nominal value, the results demonstrate that both tests can validly be used for two-sample tests with $\alpha = .05$. The close agreement of the significance level of these tests with the nominal value also allows us to fairly compare their power.

Some statisticians believe that the unequal variance t test will maintain its size better than the pooled t test. Previous published research by O'Gorman (1997b) indicates that the significance level of the unequal variance t test often exceeds 6% when used with unequal sample sizes with nonnormal error distributions. The failure of the unequal variance t test to maintain its size makes it difficult to fairly compare its power to that of other tests. Although the unequal variance t test will not be considered as a competitor to adaptive tests in this book, we will describe a two-sample test for location and scale in Chapter 8.

Table 2.3. *The empirical size in percent for the adaptive test and the pooled t test using a significance level of* $\alpha = .05$.

	Distributions								
Skewness	0	0	0	1	1	1	2	2	2
Kurtosis	1.8	3.0	6.0	4.2	5.4	8.4	11.4	12.6	15.6
	$n_1 = 6,\ n_2 = 6$								
Adaptive	5.3	4.9	5.0	4.9	4.6	4.8	5.0	5.1	5.3
t test	5.6	5.0	4.5	4.6	4.2	4.3	4.3	4.3	4.4
	$n_1 = 10, n_2 = 10$								
Adaptive	5.0	5.0	5.1	5.3	5.4	5.3	5.1	4.9	5.1
t test	5.4	5.0	4.7	5.0	5.2	4.9	4.4	4.3	4.3
	$n_1 = 20, n_2 = 20$								
Adaptive	4.8	4.9	4.9	5.2	5.2	5.1	4.8	5.3	4.8
t test	4.8	4.9	4.7	5.2	5.1	4.8	4.7	5.1	4.5
	$n_1 = 50, n_2 = 50$								
Adaptive	5.2	4.7	4.7	5.1	4.9	4.9	4.8	4.9	4.9
t test	5.3	4.8	4.3	4.9	4.6	4.8	5.1	4.9	4.6
	$n_1 = 6,\ n_2 = 25$								
Adaptive	5.6	4.7	4.8	5.0	5.4	5.0	5.3	5.0	5.3
t test	5.5	4.7	5.2	4.7	5.5	5.2	4.9	4.7	5.2
	$n_1 = 48,\ n_2 = 12$								
Adaptive	5.3	5.0	4.9	5.1	4.8	4.9	5.5	5.2	5.3
t test	5.1	4.8	4.9	5.1	4.6	5.0	5.2	4.9	5.3

The empirical sizes of the pooled t test and the proposed adaptive test were also obtained for the $\alpha = .01$ level of significance for the nine distributions and sample size configurations. To save space, these results have not been tabulated. For these estimates of size, the standard error is 0.10%, so we would expect that approximately 95% of the size estimates should fall in the range $1\% \pm 2(0.10)\% = [0.8\%, 1.2\%]$. For the adaptive test only one empirical size fell outside that range in these simulations, and it was 1.31% for the simulation having $n_1 = n_2 = 10$ with $\alpha_3 = 0$ and $\alpha_4 = 6.0$. This does not prove that the adaptive test maintains its size exactly, but does show that, for these sample sizes and error distributions, the empirical size of the adaptive test appears to be close to the nominal value. In contrast, the empirical size of the t test fell outside of the range $1\% \pm 2(0.10)\%$ for 11 of the 54 simulations. Thus, while both tests had empirical sizes close to the nominal value, the size of the adaptive test was generally closer to the nominal value.

2.7.3 The Power of the Adaptive Test

For two-sample tests, the location shift (δ) is the mean of the second population minus the mean of the first population. Figure 2.8 gives the empirical power of the proposed test for several values of δ, which were determined so that the powers of the pooled t test would be 10%, 25%, 50%, 75%, and 95%. For those simulations with $(n_1, n_2) = (30, 30)$ we used

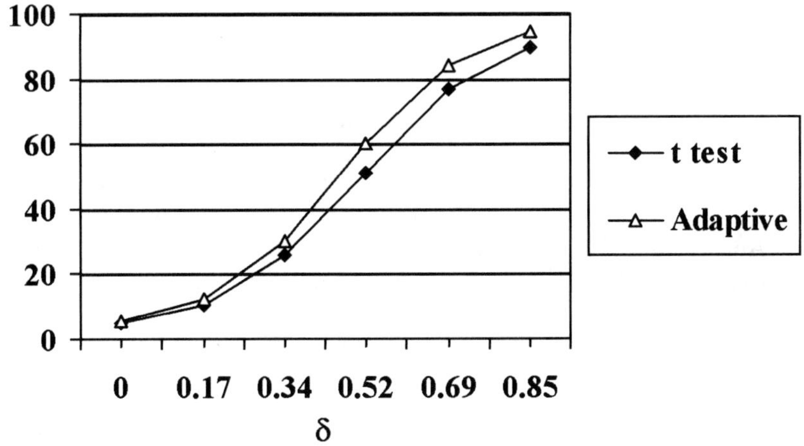

Figure 2.8. *The empirical power in percent for the adaptive test and the pooled t test using $\alpha = .05$ for two independent samples of size $n_1 = 30,\ n_2 = 30$ from a skewed distribution with $\alpha_3 = 1,\ \alpha_4 = 5.4$.*

a moderately skewed generalized lambda distribution with $\alpha_1 = 0,\ \alpha_2 = 1.0,\ \alpha_3 = 1.0$, and $\alpha_4 = 5.4$ to generate error terms for both groups. We added zero to each error generated for the first population and δ to each error generated for the second population in order to obtain the desired location shift of δ. The results show that the relative performance of the adaptive test to the pooled t test does not vary over the range of location shifts used in these simulations. Since the relative performance of the test is not too sensitive to δ, and since we are not too interested in comparing the performance of these tests for large values of δ where the powers of both tests exceed 99%, it seems reasonable to compare the relative performance of the tests by comparing their powers using a value of δ that achieves a power of approximately 50% for the pooled t test.

The empirical power estimates for the proposed adaptive test and the pooled t test are given in Table 2.4 for values of δ determined to achieve a power of approximately 50% for the pooled t test. The power results displayed in Table 2.4 show that, for nonnormal errors, the adaptive test often has a 5% to 20% power advantage over the t test. For the approximate normal distribution, which has $\alpha_3 = 0.0$ and $\alpha_4 = 3.0$, the t test was slightly more powerful than the adaptive test, but the difference in power was approximately 1%. It should be noted that for nonnormal errors the adaptive test has only a small power advantage over the pooled t test for samples with $n_1 = n_2 = 6$, but it often has much greater power than the t test for larger samples with $n_1 + n_2 \geq 20$. Thus, if $n_1 + n_2 \geq 20$, there is much to gain and little to lose if adaptive tests are used instead of t tests.

Another simulation study was performed to determine the effectiveness of the adaptive test with a significance level of $\alpha = .01$. Except for the significance level, the design of the study was identical to that used for $\alpha = .05$. The empirical power of the adaptive test and the pooled t test are tabulated in Table 2.5 for the $\alpha = .01$ level of significance. These results show the same pattern that we observed for the $\alpha = .05$ level tests. The adaptive test is usually more powerful than the t test for nonnormal error distributions if $n_1 + n_2 \geq 20$, but

Table 2.4. *The empirical power in percent for the adaptive test and the pooled t test using a significance level of* $\alpha = .05$.

	Distributions								
Skewness	0	0	0	1	1	1	2	2	2
Kurtosis	1.8	3.0	6.0	4.2	5.4	8.4	11.4	12.6	15.6
			$n_1 = 6, n_2 = 6$						
Adaptive	46.8	49.5	54.1	53.7	54.9	57.1	62.9	62.6	62.0
t test	46.9	50.9	54.4	52.3	54.2	56.4	59.5	59.7	59.4
			$n_1 = 10, n_2 = 10$						
Adaptive	51.3	48.4	55.6	55.4	56.3	57.5	67.5	67.0	65.9
t test	49.4	49.5	53.7	50.5	52.8	53.4	56.2	56.9	57.8
			$n_1 = 20, n_2 = 20$						
Adaptive	55.9	48.4	55.5	60.5	58.3	59.8	74.6	70.5	68.7
t test	49.1	49.6	51.8	50.7	51.2	52.6	54.6	53.8	54.6
			$n_1 = 50, \ n_2 = 50$						
Adaptive	65.4	48.2	56.1	65.3	60.0	61.8	78.5	75.1	71.4
t test	49.6	48.9	50.4	49.8	50.2	52.3	52.2	52.2	53.3
			$n_1 = 6, \ n_2 = 25$						
Adaptive	53.4	48.9	54.6	61.2	59.8	59.9	71.6	69.9	67.4
t test	49.5	50.3	51.6	52.9	54.2	54.4	56.5	57.2	56.7
			$n_1 = 48, \ n_2 = 12$						
Adaptive	60.1	48.1	54.8	60.2	57.3	58.5	74.3	72.3	69.3
t test	49.8	49.1	50.1	49.2	49.0	50.3	51.1	50.9	51.8

the t test is slightly more powerful than the adaptive test for small samples with $n_1 = n_2 = 6$ and for error distributions that are approximately normal. Thus, if $\alpha = .01$ and the errors are nonnormal, the adaptive test has a power advantage over the t test whenever the number of observations equals or exceeds 20.

2.7.4 The Effectiveness of the Smoothing Procedure

A small simulation study was performed to see if the procedure used to smooth the c.d.f. was really beneficial. If we did not smooth the residuals, we found that the power of the adaptive test was 4.3% below the power of the t test for $(n_1, n_2) = (30, 30)$ when the errors were from a distribution with $\alpha_3 = 0$ and $\alpha_4 = 3.0$, whereas, if we smoothed the residuals, we found that the power of the adaptive test was only 1.4% below the power of the t test. In another simulation for $(n_1, n_2) = (30, 30)$ that had errors from a distribution with $\alpha_3 = 1$ and $\alpha_4 = 5.4$ we found, if we did not smooth the residuals, that the adaptive test was 6.2% more powerful than the t test, whereas, if we smoothed the residuals, we found that the adaptive test was 8.4% more powerful than the t test. Based on these results we concluded that the smoothing of the empirical c.d.f. did increase the overall performance of the test. Of course, smoothing methods other than the one we have chosen could be used, but the chosen method does seem to be effective.

Table 2.5. *The empirical power in percent for the adaptive test and the pooled t test using a significance level of* $\alpha = .01$.

	Distributions								
Skewness	0	0	0	1	1	1	2	2	2
Kurtosis	1.8	3.0	6.0	4.2	5.4	8.4	11.4	12.6	15.6
				$n_1 = 6,\ n_2 = 6$					
Adaptive	38.6	45.9	53.5	50.5	53.1	56.2	59.7	59.1	59.7
t test	45.1	49.6	55.2	52.3	55.1	57.5	60.8	60.0	59.9
				$n_1 = 10,\ n_2 = 10$					
Adaptive	46.2	49.3	55.5	54.4	55.7	58.5	65.9	66.4	65.8
t test	46.8	50.8	54.0	51.5	53.0	56.3	57.3	58.6	59.1
				$n_1 = 20,\ n_2 = 20$					
Adaptive	54.3	48.4	55.7	61.7	59.6	61.5	75.1	74.2	70.3
t test	49.7	49.8	51.8	51.3	52.3	53.9	55.1	55.7	55.6
				$n_1 = 50,\ n_2 = 50$					
Adaptive	66.5	47.8	56.9	68.1	62.2	53.1	83.8	80.4	75.8
t test	50.2	49.1	50.1	50.3	50.7	51.7	52.7	52.8	53.8
				$n_1 = 6,\ n_2 = 25$					
Adaptive	48.9	48.6	55.3	60.1	59.8	60.6	72.6	69.6	69.4
t test	48.3	50.7	51.9	52.4	54.1	55.1	58.4	56.3	58.2
				$n_1 = 48, n_2 = 12$					
Adaptive	57.6	47.7	57.2	59.3	57.8	60.9	76.3	75.1	71.9
t test	48.6	49.4	52.2	48.8	50.0	52.0	51.2	53.5	53.6

2.8 A Comparison of the Adaptive Test to the Wilcoxon Rank-Sum Test

We have seen that the adaptive test is usually more powerful than the *t* test for nonnormal errors for samples with $n_1 + n_2 \geq 20$. We now turn our attention to the Wilcoxon rank-sum test, which is another popular two-sample test. The Wilcoxon test is based on the sum of the ranks in the second sample and can be computed using the Wilcoxon scores and the test statistic described in Chapter 1. The Wilcoxon test is not too sensitive to outliers and is the locally most powerful rank test for detecting a shift in a logistic distribution (see Randles and Wolfe (1979, Chap. 9)). The simulation study performed to compare the adaptive test to the *t* test also included the Wilcoxon test. We found, as expected, that the Wilcoxon test maintained its level of significance for the nine distributions and six sample size configurations that we used in the simulation studies.

The power of the adaptive test and the Wilcoxon test is displayed in Table 2.6. We observe little difference in power between these tests for samples with $n_1 = n_2 = 10$, but the adaptive test is sometimes more powerful than the Wilcoxon test for $n_1 = n_2 = 20$ and $n_1 = n_2 = 50$. It is clear from these results and from those in Table 2.4 that both the adaptive test and the Wilcoxon rank-sum test are considerably more powerful than the *t* test

Table 2.6. *The empirical power in percent for the adaptive test and the Wilcoxon rank-sum test using a significance level of* $\alpha = .05$.

	Distributions								
Skewness	0	0	0	1	1	1	2	2	2
Kurtosis	1.8	3.0	6.0	4.2	5.4	8.4	11.4	12.6	15.6
	$n_1 = 10, n_2 = 10$								
Adaptive	51.3	48.4	55.6	55.4	56.3	57.5	67.5	67.0	65.9
Wilcoxon	46.0	48.3	57.7	54.7	56.2	58.9	67.0	66.4	66.4
	$n_1 = 20, n_2 = 20$								
Adaptive	55.9	48.4	55.5	60.5	58.3	59.8	74.6	70.5	68.7
Wilcoxon	45.3	47.3	57.2	55.7	56.8	61.0	70.5	67.3	67.0
	$n_1 = 50, n_2 = 50$								
Adaptive	65.4	48.2	56.1	65.3	60.0	61.8	78.5	75.1	71.4
Wilcoxon	46.4	47.5	57.7	57.3	57.6	62.3	72.0	70.2	69.4

for many distributions. The Wilcoxon test is a simple and effective two-sample test but is often not as powerful as the adaptive test, nor can it be generalized so that it can be used for more complex models.

2.9 Concluding Remarks on Two-Sample Tests

The adaptive test appears to be robust for validity because simulation studies have shown that it maintains its size close to the nominal value. Often, the empirical size of the adaptive test is closer to the nominal value than the empirical size of the t test. Furthermore, the simulation results demonstrate that the adaptive test is often more powerful than the t test. The power difference between the tests can sometimes be substantial; for skewed error distributions the power differences are usually between 5% and 20%. Conversely, the power loss observed with the adaptive test is small for errors with the approximate normal distribution. For the simulations that used $\alpha = .05$ the power advantage of the t test never exceeded 1.5%. Consequently, the potential advantage in power of the adaptive test with nonnormal distributions greatly exceeds the potential disadvantage with the normal distribution.

We have also compared the power of the Wilcoxon test to that of the adaptive test and have found little difference in power between them for small sample sizes. For moderate and large samples with 40 or more observations the adaptive test is sometimes more powerful than the Wilcoxon test. The adaptive test seems to be preferable to the Wilcoxon test because it can be applied in many more testing situations than the Wilcoxon test and usually is at least as effective. The simulation results also show that the t test is not a real competitor to the adaptive and Wilcoxon tests.

The adaptive test does require special software. However, the software, which is available in the form of a SAS macro, is easy to use and executes in a reasonable amount of time on a modern computer. This software is described in Chapter 4, where it is used to analyze several kinds of data.

In the development of the weighting scheme we compared the smoothed c.d.f. of the studentized deleted residuals to the c.d.f. of the t distribution. We could have used probability plots instead of distribution functions to determine how closely the residuals followed the t distribution. This was the approach taken by O'Gorman (2001b) to construct a weighting procedure for adaptive confidence intervals. The major problem with the probability plot approach is that it is difficult to find a good procedure to smooth the points on a probability plot, whereas smoothing procedures for distribution functions have already been developed. In addition, simulation results have shown that the test that uses the distribution function approach is slightly more powerful than the test that uses the probability plot approach.

We should also keep in mind that the weighting procedure may not need to be absolutely perfect in order to be effective. In a typical research situation we do not know how the errors are distributed, so it may not be possible to accurately determine the correct weights. Further, because the adaptive test is performed using WLS, it is the relative weights that are important, not the absolute weights. This has important consequences for the construction of weighting procedures because consistently under- or overestimating the standard error of the deleted residuals by a factor will have the effect of increasing or decreasing the weights by a constant, so that the adaptive test will be unchanged.

Chapter 3

A General Adaptive Testing Method

3.1 An Adaptive Test for a Subset of Regression Coefficients

The results in the last chapter demonstrated the superiority of the adaptive approach for two-sample tests when the total sample size equaled or exceeded 20. In this chapter we will describe a general approach to adaptive testing that is similar to the approach used with the two-sample tests. In our general approach we will adaptively weight the observations and then use a permutation method to perform the test. Although the weighting procedure is a straightforward generalization of the procedure used with the two-sample tests, the permutation method requires further explanation. In this chapter the general adaptive testing procedure, which can be used with complex models, will be described in detail.

In order to illustrate the general adaptive testing method we will use data collected by Haith (1976) on water quality and land use in 20 river basins in New York State. These data, which are listed in Table 3.1, have been analyzed by several authors, including Simpson et al. (1992) and Ryan (1997). The data include the nitrogen concentration (mg/L) of river water and several land use variables. The nitrogen concentration will be used as the dependent variable. The independent variables are the percentages of commercial, agricultural, forest, and residential land in the river basin. Because residential land use is highly correlated with commercial land use we will not include the former in our model. Thus, for this analysis, we will use the multiple regression model

$$y_i = \beta_0 + \beta_1 x_{i,1} + \beta_2 x_{i,2} + \beta_3 x_{i,3} + \varepsilon_i,$$

where $x_{i,1}$ is the percentage of commercial land, $x_{i,2}$ is the percentage of agricultural land, and $x_{i,3}$ is the percentage of forested land.

Suppose we have already been convinced that commercial and agricultural land are predictive of nitrogen concentration, but we are uncertain about the influence of forested land. To investigate this relationship we test $H_0 : \beta_3 = 0$ against $H_a : \beta_3 \neq 0$. The traditional F test gives a test statistic of $F = 1.98$, which yields a p-value of $p = .179$. If we used a significance level of $\alpha = .10$ we would not reject the null hypothesis. Consequently, based on this traditional analysis, we cannot say that the percentage of forested land is

Table 3.1. *The New York rivers data set.*

	River Basin	Y Nitrogen	X_1 Commercial	X_2 Agric.	X_3 Forest	X_4 Res.
1	Olean	1.10	0.29	26	63	1.2
2	Cassadaga	1.01	0.09	29	57	0.7
3	Oatka	1.90	0.58	54	26	1.8
4	Neversink	1.00	1.98	2	84	1.9
5	Hackensack	1.99	3.11	3	27	29.4
6	Wappinger	1.42	0.56	19	61	3.4
7	Fishkill	2.04	1.11	16	60	5.6
8	Honeoye	1.65	0.24	40	43	1.3
9	Susquehanna	1.01	0.15	28	62	1.1
10	Chenango	1.21	0.23	26	60	0.9
11	Tioughnioga	1.33	0.18	26	53	0.9
12	West Canada	0.75	0.16	15	75	0.7
13	East Canada	0.73	0.12	6	84	0.5
14	Saranac	0.80	0.35	3	81	0.8
15	Ausable	0.76	0.35	2	89	0.7
16	Black	0.87	0.15	6	82	0.5
17	Schohari	0.80	0.22	22	70	0.9
18	Raquette	0.87	0.18	4	75	0.4
19	Oswegatchie	0.66	0.13	21	56	0.5
20	Cohocton	1.25	0.13	40	49	1.1

related to nitrogen concentration, after controlling for the percentage of commercial and agricultural land.

Before we perform an adaptive test of $H_0 : \beta_3 = 0$ against $H_a : \beta_3 \neq 0$ in this multiple regression model, we will describe a general adaptive testing procedure that can be used for any subset of regression parameters. We will assume that we have n observations and that for each observation there are $p + 1$ independent variables. The linear model that we will use is

$$y_i = \beta_0 x_{i,0} + \beta_1 x_{i,1} + \cdots + \beta_p x_{i,p} + \varepsilon_i$$

for $i = 1, \ldots, n$. Note that, in order to incorporate an intercept term into the model, we will often set $x_{i,0} = 1$ for $i = 1, \ldots, n$. We will let Y be the $n \times 1$ vector containing the dependent variable

$$Y = \begin{bmatrix} y_1 \\ \cdot \\ \cdot \\ \cdot \\ y_n \end{bmatrix}$$

and will let the independent variables for the complete model be written as the $n \times (p + 1)$ matrix

$$X_c = \begin{bmatrix} x_{1,0} & \cdot & \cdot & \cdot & x_{1,p} \\ \cdot & \cdot & \cdot & \cdot & \cdot \\ \cdot & \cdot & \cdot & \cdot & \cdot \\ \cdot & \cdot & \cdot & \cdot & \cdot \\ x_{n,0} & \cdot & \cdot & \cdot & x_{n,p} \end{bmatrix}.$$

The $p + 1$ parameters in the complete model can be written as a $(p + 1) \times 1$ column vector

$$\beta_C = \begin{bmatrix} \beta_0 \\ \beta_1 \\ \cdot \\ \cdot \\ \beta_p \end{bmatrix},$$

and the error term in the model can be written as the $n \times 1$ column vector

$$\varepsilon_C = \begin{bmatrix} \varepsilon_1 \\ \cdot \\ \cdot \\ \cdot \\ \varepsilon_n \end{bmatrix}.$$

With these definitions the complete model can be written in matrix form as

$$Y = X_C \beta_C + \varepsilon_C.$$

Now consider the reduced model with q independent variables:

$$y_i = \beta_0 x_{i,0} + \beta_1 x_{i,1} + \cdots + \beta_q x_{i,q} + \varepsilon_i,$$

where $0 \le q < p$. The reduced model can also be written in matrix form as

$$Y = X_R \beta_R + \varepsilon_R,$$

where X_R is an $n \times (q + 1)$ matrix that contains the first $q + 1$ columns of X_C, β_R is a $(q + 1) \times 1$ vector of parameters, and ε_R is an $n \times 1$ vector of errors.

We often need to determine if the $p - q$ additional parameters that are in β_C, but are not in β_R, are needed to adequately explain Y. The F test for a subset of parameters in a linear model is traditionally used for this purpose. Let β_A be the vector containing the last $p - q$ elements in β_C. We usually test $H_0 : \beta_A = 0$ versus $H_a : \beta_A \neq 0$ with the F test, which requires fitting the complete and reduced models and comparing the fit of both models. The traditional test statistic is

$$F = \frac{(SSE_R - SSE_C)/(p - q)}{SSE_C/(n - p)},$$

where SSE_R is the sum of squared errors for the reduced model and SSE_C is the sum of squared errors for the complete model. We reject $H_0 : \beta_A = 0$ at the α level of significance if $F \ge F_{\alpha,p-q,n-p-1}$, where F_{α,ν_1,ν_2} is the critical value of the F distribution with ν_1 and ν_2 degrees of freedom.

The F test is robust for validity and is known to be the most powerful test if the errors are normally distributed, are independent, and have constant variance. Further, the F test is easy to compute because software for computing the sum of squared errors is commonly available. However, if the errors are not normally distributed, then the F test may not be the most powerful test. We turn now to an adaptive test that is also robust for validity and often has greater power than the F test if the errors are not normal.

3.2 Computing and Smoothing Residuals

For this general adaptive test we will use the studentized deleted residuals to weight the observations in the same way as they were used in the two-sample test. For the general linear model Belsley, Kuh, and Welsch (1980) express the studentized deleted residuals for the ith observation in the reduced model as

$$d_i = \frac{e_i}{s(i)\sqrt{1 - h_{ii}}},$$

where e_i is the ordinary residual, h_{ii} is the ith diagonal element of the hat matrix $X_R(X_R'X_R)^{-1}X_R'$, and $s(i)$ is the estimate of σ based on the $n-1$ observations obtained by deleting the ith observation from the data set.

Belsley, Kuh, and Welsch (1980) then show that $s(i)$ can be computed without recomputing the least squares estimates by using the relationship

$$(n - q - 2)s^2(i) = SSE_R - \frac{e_i^2}{1 - h_{ii}},$$

where SSE_R is the usual sum of the squared residuals from the regression based on the n observations in the reduced model. It follows that the studentized deleted residual for the ith observation is

$$d_i = e_i \left[\frac{n - q - 2}{SSE_R\,(1 - h_{ii}) - e_i^2} \right]^{\frac{1}{2}}$$

for $i = 1, \ldots, n$. If the errors are normally distributed, each studentized deleted residual is distributed as Student's t with $\nu = (n-1) - (q+1) = n - q - 2$ degrees of freedom. However, the studentized residuals are not independent. The number of degrees of freedom is based on the fact that only $n-1$ of the observations are used in computing deleted residuals for the reduced model with $q+1$ independent variables.

As in the previous chapter, the observations will be weighted so that the c.d.f. of the studentized deleted residuals, after weighting, will approximate the c.d.f. of the t distribution with $\nu = n - q - 2$ degrees of freedom. This weighting is done to normalize the errors, much in the way a transformation would be used to normalize the errors. However, we weight each observation, rather than transform the y values, as would be done with a Box–Cox transformation. The advantage of this approach is that we continue to use the linear relationship between the independent variables and the dependent variable that was in the original linear model.

In order to increase the effectiveness of the testing procedure, the c.d.f. of the studentized deleted residuals will be smoothed using a normal kernel with a bandwidth of

$h = 1.587\,\hat{\sigma}\, n^{-1/3}$, as suggested by Polansky (1998). This is the same smoothing procedure as we used for the two-sample adaptive test. The smoothing method is used in order to provide a better estimate of the weights to be given to each observation. Since the deleted residuals are studentized the variance should not depart too much from $\sigma^2 = 1$, so a value of $h = 1.587\ n^{-1/3}$ is used to obtain the smoothed distribution function. Let $D = \{d_1, \ldots, d_n\}$ be the set of studentized deleted residuals. The smoothed c.d.f. at point d over the set of all studentized deleted residuals (D) is computed as

$$\hat{F}_h(d; D) = \frac{1}{n} \sum_{i=1}^{n} \Phi\left(\frac{d - d_i}{h}\right),$$

where $\Phi(\cdot)$ is the c.d.f. of the standard normal distribution. For $n = 32$ the bandwidth is $h = 1.587(32)^{-1/3} = 0.5$. For larger values of n we obtain smaller values for the bandwidth, which means that the smoothing is more local.

3.3 Weighting Observations

After we obtain the studentized deleted residuals we center them by subtracting the estimated median ($\tilde{d}$), which is determined by a root-finding algorithm so that $\hat{F}_h(\tilde{d}; D) = .5$. Thus, we calculate the centered studentized deleted residuals as $d_{c,i} = d_i - \tilde{d}$ for $i = 1, \ldots, n$. For brevity, the centered studentized deleted residuals will be called the residuals. Let $D_c = \{d_{c,1}, \ldots, d_{c,n}\}$ be the set of residuals and let $t_i = T^{-1}_{n-q-2}(\hat{F}_h(d_{c,i}; D_c))$ be the inverse t value with $\nu = n - q - 2$ degrees of freedom for the ith observation with a smoothed c.d.f. of $\hat{F}_h(d_{c,i}; D_c)$. To weight the observations we use

$$w_i = \frac{t_i}{d_{c,i}} = \frac{T^{-1}_{n-q-2}(\hat{F}_h(d_{c,i}; D_c))}{d_{c,i}}$$

for $i = 1, \ldots, n$. If the smoothed c.d.f. of the residuals approximates the c.d.f. of the t distribution, then the weights will be near 1.0 and the result from the adaptive test should be close to that obtained by the traditional test. In general, observations with extremely large or small residuals will be given weights less than 1.0. If the distribution of residuals has short tails, then the observations with residuals near its ends will be given weights larger than 1.0.

An example may illustrate the rationale for using the weighting scheme. Suppose we want to perform a test for slope in a simple linear regression with an outlier data set having $n = 100$ observations. If we tested $H_0 : \beta_1 = 0$ against $H_1 : \beta_1 \neq 0$ with a model $y_i = \beta_0 + \beta_1 x_i + \varepsilon_i$, we would use the reduced model $y_i = \beta_0 + \varepsilon_i$ to compute the studentized deleted residuals. Figure 3.1 gives the histogram for the studentized deleted residuals under the reduced model for some simulated data, which included 99 of these observations generated from a normal distribution with $\mu = 4$ and $\sigma = 1$, and one outlier that was set equal to 10. The outlier, which had a deleted studentized residual of $d_i = 6.428$, is clearly shown in Figure 3.1. Note that the studentized deleted residuals are highly skewed in this data set. The outlier was given a weight of $w_i = 0.370$, while the weights for the 99 other observations fell in the narrow range between 1.02 and 1.18. Consequently, the outlier has been downweighted. As we indicated at the end of the last chapter, the relative weights are important, not the raw weights.

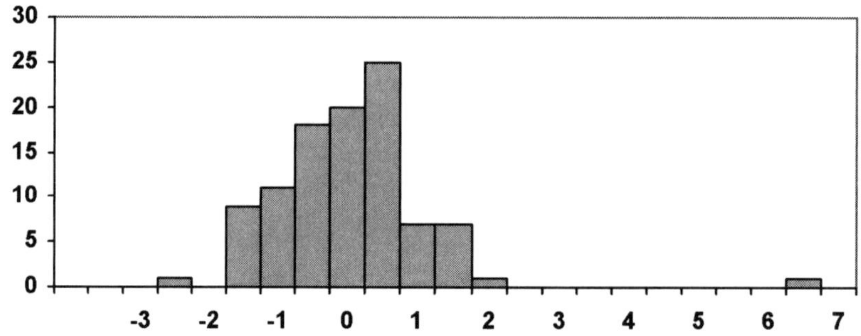

Figure 3.1. *Histogram of the studentized deleted residuals for a simulated data set.*

In the general linear model the adaptive weighting procedure has the potential of increasing the efficiency of tests of significance. After we have computed the weights w_i for $i = 1, \dots, n$, they can be used as the diagonal elements in a weighting matrix W with zero off-diagonal elements. We perform the WLS regression by premultiplying both sides of the model by W to obtain

$$WY = WX_C\beta_C + W\varepsilon_C.$$

This WLS regression model can be written as the transformed complete model

$$Y^* = X_C^*\beta_C + \varepsilon_C^*,$$

where $Y^* = WY$, $X_C^* = WX_C$, and $\varepsilon_C^* = W\varepsilon_C$. Using the transformed data, OLS methods are used to compute the sum of squared errors, which will be denoted by SSE_C^*. We can also premultiply the reduced model by W to obtain the transformed reduced model

$$Y^* = X_R^*\beta_R + \varepsilon_R^*,$$

where $X_R^* = WX_R$ and $\varepsilon_R^* = W\varepsilon_R$. We then use OLS to compute the sum of squared errors for the reduced model, which will be denoted by SSE_R^*.

We begin our adaptive test by using SSE_R^* and SSE_C^* to compute the test statistic

$$F^* = \frac{(SSE_R^* - SSE_C^*)/(p-q)}{SSE_C^*/(n-p)},$$

which is used in the test of $H_0 : \beta_A = 0$ versus $H_a : \beta_A \neq 0$. Because the adaptive test uses the data to modify the test procedure, the F^* statistic does not follow the F distribution, even if the usual inference assumptions are satisfied. Thus, if we compute the p-value directly from the F distribution and then reject the null hypothesis if that p-value is less than α, the resulting test will not have a significance level near α, so it will not be a valid test. In the next section we will describe a permutation method for the general linear model that does keep the actual significance level close to the nominal value.

In order to demonstrate the effectiveness of the weighting procedures on the simulated data, we can examine the residuals from the weighted model. After the observations are

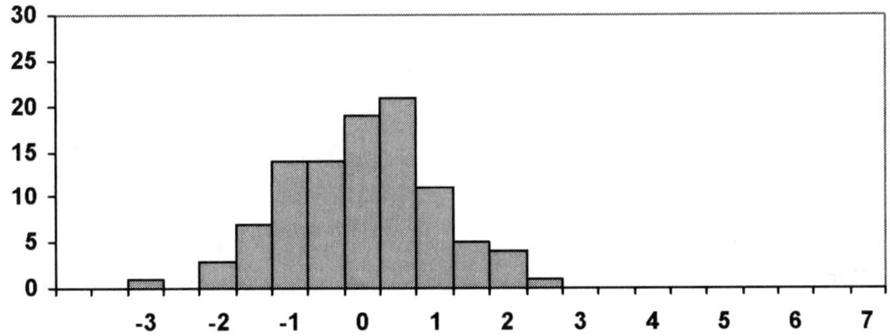

Figure 3.2. *Histogram of the studentized deleted residuals for the transformed reduced model.*

weighted, we obtain studentized deleted residuals that appear to be roughly normal, as indicated in Figure 3.2. After weighting, the studentized deleted residual for the outlier value was 2.219. This shows that, in this data set, the adaptive scheme was effective in normalizing the errors.

3.4 A General Permutation Technique

In order to describe this permutation test, we first partition the complete design matrix into the first $q+1$ columns specified in the reduced model and the additional $p-q$ columns needed for the complete model. Thus, the complete design matrix is

$$X_C = [X_R \,|X_A\,],$$

where X_A is the matrix containing the additional $p-q$ columns. If we partition β into β_R and β_A, the complete model can be written as

$$Y = [X_R|X_A]\begin{bmatrix} \beta_R \\ \beta_A \end{bmatrix} + \varepsilon_C.$$

In the permutation test of $H_0 : \beta_A = 0$ against $H_a : \beta_A \neq 0$, each permutation replication is obtained by permuting the rows of X_A. The rows in X_R and the observations in Y are not permuted. This permutation method was investigated by Kennedy and Cade (1996), who recommended using a pivotal test statistic in the permutation test. Since the weights are determined from the reduced model $Y = X_R\beta_R + \varepsilon_R$, the weights associated with the observations are not permuted, so that SSE_R^* will not change with permutations of X_A. If we let $X_{A,i}$ represent the ith permutation of the rows of X_A and let $X_{C,i}$ represent the complete design matrix corresponding to the ith permutation of the rows of X_A, the complete model corresponding to the ith permutation can be expressed as

$$Y = \left[X_R|X_{A,i}\right]\begin{bmatrix} \beta_R \\ \beta_{A,i} \end{bmatrix} + \varepsilon_C = X_{C,i}\beta_i + \varepsilon_C,$$

where β_i is the vector of parameters for the ith permutation. After the weights, which were computed from the reduced model, are applied to the rows in the model corresponding to the ith permutation, we obtain

$$WY = WX_{C,i}\beta_i + W\varepsilon_C,$$

which can be written as

$$Y^* = X^*_{C,i}\beta_i + \varepsilon^*_C,$$

where $X^*_{C,i} = WX_{C,i}$. Using this model, we compute the sum of squared residuals as $SSE^*_{C,i}$ and compute the usual F test statistic for a portion of the transformed model as

$$F^*_i = \frac{(SSE^*_R - SSE^*_{C,i})/(p-q)}{SSE^*_{C,i}/(n-p-1)}.$$

As we indicated in Chapter 2, for most applications it is not practical to compute the full permutation distribution using $n!$ permutations, so we usually use 2000 to 10,000 permutations of the rows of X_A to estimate the observed significance level. If we let E equal the number of permutations with permutation test statistics that equal or exceed the observed test statistic F^* and let R equal the total number of permutations, we can compute the p-values using the formula $p = (E+1)/(R+1)$. As usual, we reject $H_0 : \beta_A = 0$ if the p-value is less than or equal to α. It should be noted that, because SSE^*_R does not change with each permutation of the last $p - q$ columns, we could also define E to be the number of times that $SSE^*_{C,i}$ is less than or equal to SSE^*_C.

It may seem inappropriate to use the reduced model, rather than the complete model, to compute the weights. However, by using the reduced model, we use the same weights for all permutations, including the observed design matrix. Thus, the formula $p = (E+1)$ / $(R+1)$ can be validly used because the observed test statistic F^* is computed from one of the $R+1$ equally likely permutations of the rows of X_A.

3.5 An Adaptive Test for One Parameter in a Regression Model

We begin our analysis of the New York rivers data given in Table 3.1. We want to test $H_0 : \beta_3 = 0$ against $H_a : \beta_3 \neq 0$ in the complete model

$$y_i = \beta_0 + \beta_1 x_{i,1} + \beta_2 x_{i,2} + \beta_3 x_{i,3} + \varepsilon_i,$$

where y_i is the nitrogen concentration, $x_{i,1}$ is the percentage of commercial land, $x_{i,2}$ is the percentage of agricultural land, and $x_{i,3}$ is the percentage of forested land in the ith river basin. For our test of $H_0 : \beta_3 = 0$ we use the reduced model

$$y_i = \beta_0 + \beta_1 x_{i,1} + \beta_2 x_{i,2} + \varepsilon_i$$

to compute the appropriate weights for the observations. In this example Y is the 20×1 vector of nitrogen concentration and X_R is a 20×3 matrix containing an intercept indicator,

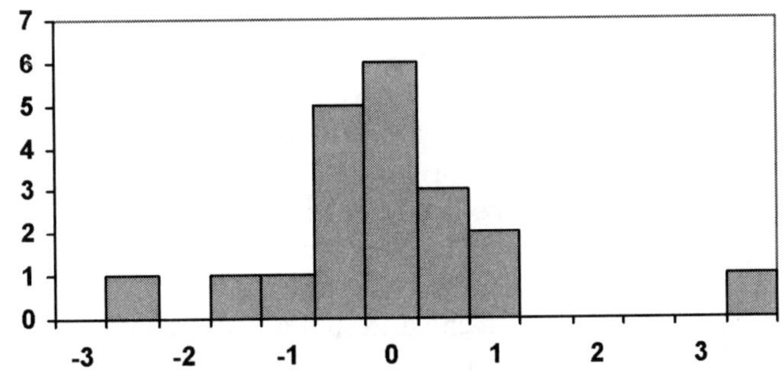

Figure 3.3. *Histogram of the studentized deleted residuals for the reduced model using the New York rivers data set.*

the percentage of commercial land, and the percentage of agricultural land. The matrix X_A is the 20×1 matrix containing the percentage of forested land. We use the reduced model $Y = X_R\beta_R + \varepsilon_R$, which has $q = 2$, to compute the deleted studentized residuals. A histogram of these deleted studentized residuals, which is presented in Figure 3.3, indicates that the residuals are slightly skewed to the right.

The smoothed and centered studentized residuals are shown in Figure 3.4 along with the c.d.f. of the t distribution with $\nu = n - q - 2 = 20 - 2 - 2 = 16$ degrees of freedom. From this figure we can see that most of the weights will be near one because the smoothed and centered studentized deleted residuals closely agree with the c.d.f. of the t distribution. However, the data from the Fishkill river basin (observation 7) has $d_7 = 3.50$

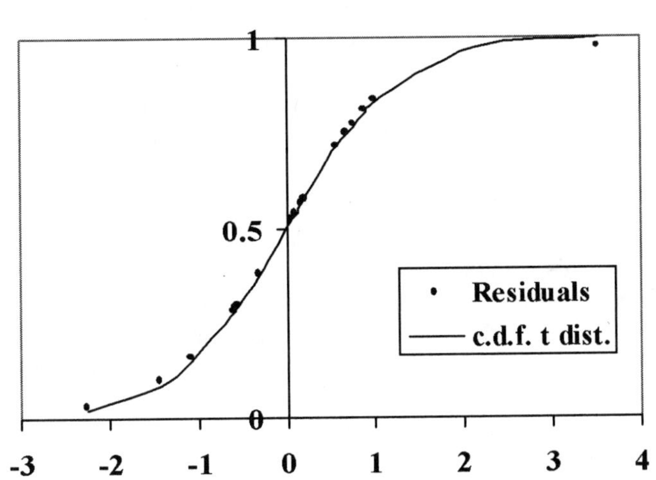

Figure 3.4. *The smoothed and centered c.d.f. of the studentized deleted residuals for the New York rivers data set and the c.d.f. of the t distribution with* $\nu = n - q - 2 = 16$ *degrees of freedom.*

and $t_7 = 2.12$. Thus, it is given a weight of $w_7 = 0.602$, while the other weights are in the interval [0.89, 1.03]. Consequently, the data from the Fishkill river basin will have less influence on the test than they would have had they been used in an ordinary regression model. It should be noted that the Hackensack river basin, which has the highest proportion of commercial land, is given a weight of $w_5 = 1.005$ because the centered studentized deleted residual of $d_{c,5} = 0.09434$ closely agrees with the t inverse value of $t_5 = 0.09480$. This demonstrates that observations that are outliers in the independent variables are not always downweighted by the adaptive methods.

We then used the permutation method to shuffle the rows of the X_A matrix, which consisted of a single column containing the forest land use data. The rows in X_R, which contains the indicator for the intercept and the data for commercial and agricultural land use, are not permuted. We used $R = 2000$ permutations of X_A and obtained a p-value for the adaptive test of $p = .0665$. Although this p-value exceeds the commonly used significance level of $\alpha = .05$, it does suggest that forest land use may be related to nitrogen concentration after controlling for commercial and agricultural land use. Since most of the weights were near one, the reason for the difference between the traditional p-value, which was $p = 0.179$, and the adaptive p-value is due to the downweighting of the data from the Fishkill river basin. While it is curious to note the difference in the results obtained by these two methods, we cannot conclude that one method is superior to the other because we do not know the true relationship between nitrogen concentration and forested land use. We will now investigate the effectiveness of the general adaptive test in a multiple regression model.

3.6 The Performance of the General Adaptive Test

In our example in the last section we tested a single slope in a multiple regression model with three independent variables. We found that, if we had used $\alpha = .10$, we would have rejected the null hypothesis with the adaptive test but not with the usual F test.

In this section we will use a simulation study to investigate the size and power of a test of a single slope in a multiple regression with two independent variables. That is, we will evaluate the test of $H_0 : \beta_2 = 0$ against $H_a : \beta_2 \neq 0$ in the model

$$y_i = \beta_0 + \beta_1 x_{i,1} + \beta_2 x_{i,2} + \varepsilon_i .$$

The error terms used in this simulation are from the same nine generalized lambda distributions that were used in the simulations described in the last chapter. However, for these simulations the two independent variables were generated from a bivariate lognormal distribution and then used in the regression model along with the error terms to generate the Y values. We used the bivariate lognormal distribution because we could generate skewed variates with any specified correlation between the two independent variables. The generation of bivariate lognormal variables is discussed in Johnson (1987, Chap. 5).

For the simulations designed to evaluate the size of the tests, we set $\beta_2 = 0$ when we generated the dependent variables. For each data set generated we determined if the null hypothesis $H_0 : \beta_2 = 0$ was rejected with the adaptive test and with the traditional F test. In these simulations we used 10,000 data sets with $R = 2000$ permutations for each

Table 3.2. *The empirical size in percent for a test of* $H_0 : \beta_2 = 0$ *against* $H_a : \beta_2 \neq 0$ *using a significance level of* $\alpha = .05$.

	Distributions								
Skewness	0	0	0	1	1	1	2	2	2
Kurtosis	1.8	3.0	6.0	4.2	5.4	8.4	11.4	12.6	15.6
					$n = 10$				
Adaptive	4.9	5.1	5.2	4.6	5.2	5.2	5.4	5.0	4.9
F test	4.3	5.3	5.5	4.9	5.1	5.4	5.6	5.2	5.7
					$n = 20$				
Adaptive	5.3	5.0	4.9	5.1	5.1	5.0	4.5	5.3	5.1
F test	4.7	5.1	5.0	5.0	5.3	5.7	5.0	5.5	5.4
					$n = 40$				
Adaptive	5.0	5.5	5.0	5.3	5.0	5.0	4.9	5.3	5.2
F test	4.6	5.5	5.3	5.0	5.1	5.5	5.1	5.2	5.1
					$n = 100$				
Adaptive	5.5	5.0	5.0	4.8	4.9	5.1	4.8	5.2	5.0
F test	5.1	5.0	5.1	4.9	5.0	5.3	5.0	5.0	5.1

set. The results displayed in Table 3.2 are from simulations that used independent variables generated from a bivariate lognormal distribution with a correlation of $\rho = 0.4$. We used a moderate positive correlation because we were interested in seeing if the permutation method would be effective with correlated covariates. The results in Table 3.2 show close agreement between the empirical sizes of both tests and the nominal value of 5%.

Since we generated the data sets from a bivariate lognormal distribution with a correlation of $\rho = 0.4$ we would find, in most data sets, a moderate correlation between the independent variables. However, the correlation would not generally be observed after the rows of X_A are permuted. Even though the permutation method destroys the correlation between x_1 and x_2, we do find close agreement of these empirical sizes with the nominal value. Another simulation was performed that was identical to the one described except the correlation between the independent variables was $\rho = 0.8$. The empirical sizes of the adaptive test for these data sets ranged from 4.5% to 5.5%. Thus, the adaptive test maintained its size, even with a large correlation between the independent variables.

The empirical powers of the adaptive test are given in Table 3.3 for a simulation study that used a correlation of $\rho = 0.4$ between the independent variables. In order to obtain empirical powers in the range [.25, .75], we set $\beta_2 = 2.5/\sqrt{n}$ when we generated the dependent variable for the simulations and then counted the percentage of data sets that led to rejection of the null hypothesis $H_0 : \beta_2 = 0$. These results show that the adaptive test is more powerful than the traditional test for sample sizes of $n \geq 40$ with nonnormal errors. For simulations with errors generated from the approximately normal distribution, the traditional test was only slightly more powerful for $n \geq 20$, but somewhat more powerful for $n = 10$. Thus, the adaptive test is not recommended for $n = 10$. These results for $n = 10$ are not too surprising because there is not enough information in $n = 10$ residuals to effectively determine the appropriate weights.

Table 3.3. *The empirical power in percent for a test of $H_0 : \beta_2 = 0$ against $H_a : \beta_2 \neq 0$ using a significance level of $\alpha = .05$.*

	Distributions								
Skewness	0	0	0	1	1	1	2	2	2
Kurtosis	1.8	3.0	6.0	4.2	5.4	8.4	11.4	12.6	15.6
					$n = 10$				
Adaptive	26.4	26.6	31.0	29.1	30.1	32.5	34.6	33.3	37.1
F test	28.8	30.1	33.4	31.6	32.9	35.4	36.1	35.2	35.3
					$n = 20$				
Adaptive	40.6	39.4	43.5	43.0	43.1	44.8	50.9	50.4	48.5
F test	40.7	41.9	43.8	41.6	43.3	44.3	45.5	45.1	45.2
					$n = 40$				
Adaptive	52.0	47.4	53.7	54.0	53.2	54.9	63.7	62.8	61.5
F test	48.3	48.8	51.2	49.0	49.4	51.0	50.4	51.7	51.6
					$n = 100$				
Adaptive	66.4	54.1	60.7	66.0	62.9	64.4	77.3	75.0	72.6
F test	55.5	55.3	55.7	55.2	55.3	55.6	55.3	56.5	57.4

3.7 Comments on the Performance of the General Adaptive Test

While it is interesting to know that the adaptive test will maintain its size with regression problems having two independent variables, we cannot be certain that it will maintain its size for tests of subsets of regression coefficients for more complex models. The permutation method that we used was investigated by Kennedy and Cade (1996) for a nonadaptive test of subsets of regression coefficients in linear models. In their work the permutation approach was called the shuffle Z method.

If there is only one column in the X_A matrix the shuffle Z method appears to be related to the Manly (1997) method of permuting the raw data (Y). Let Z be the variable in the X_A matrix and let X indicate the variables in the X_R matrix. To test $H_0 : \beta_A = 0$ we want to determine if there is a relationship between Y and Z after controlling for X. Let $r_{YZ \cdot X}$ denote the partial correlation between Y and Z after controlling for the variables in the X matrix. The t statistic that is used to test $H_0 : \beta_A = 0$ is a function of $r_{YZ \cdot X}$ and n since the t statistic can be expressed as

$$t = r_{YZ \cdot X} \frac{\sqrt{n - q - 2}}{\sqrt{1 - r_{YZ \cdot X}^2}},$$

where q is the number of independent variables in X. (See, for example, Kleinbaum et al. (1998, p. 176).) When there is only one column in the X_A matrix this t test is equivalent to an F test because $F = t^2$ when $p - q = 1$. We recall that when correlations are used to compute the test statistic we do not need to make a distinction between independent and dependent variables. Consequently, we will obtain the same F test statistic if we use Z as the dependent variable and Y as the variable in X_A, since $r_{YZ \cdot X} = r_{ZY \cdot X}$ and F is a function of the partial correlation.

Hence, if we make Z the dependent variable and use Manly's (1997) method of permuting the raw data in the Y vector, we will obtain the same permutation distribution of F as we obtained with the shuffle Z method. The Type I error of the Manly method of permuting raw data has been investigated by Anderson and Legendre (1999). In their simulation studies they found that the Type I error of Manly's method did not differ significantly from $\alpha = .05$ for normal and exponential errors. However, they did find that the Type I error fell into the range $.03 - .04$ if the errors were generated from an exponential cubed distribution. They also found that the Manly test did not maintain its size near the nominal value when an outlier was present in the covariate. Anderson and Legendre (1999) generated data by choosing $n - 1$ values of a covariate from a uniform distribution on the interval (0, 3) with the remaining outlier set equal to 55. The failure of Manly's method to maintain its size was pointed out earlier by Kennedy and Cade (1996), who used the same method of generating outliers for the covariate. Although the matter is not completely settled, it seems clear that Manly's method, which uses no adaptation, maintains its size near the nominal value except when extreme situations are simulated. This view is also expressed by Manly (1997, sect. 8.6).

When adaptive methods are used, the investigation of the size of the adaptive test appears to be more complex. The simulation results shown in this chapter and the next document the close agreement of the empirical significance level of the adaptive test with the nominal significance level. Based on these simulation studies the author has concluded that the adaptive test usually has a size that is close to the nominal significance level of α. However, we have not attempted to simulate situations that are as extreme as those used in Anderson and Legendre (1999).

The fact that the adaptive test does not maintain the exact significance level is not a very important practical problem because the departure from the exact significance level does not appear to be large. This is shown in Table 3.2, where it can be seen that the empirical size of the adaptive test is close to α for all distributions. It should also be remembered that the traditional t and F tests do not maintain their size for nonnormal errors. Indeed, for small sample sizes the empirical size of the adaptive test is often much closer to the nominal value than the empirical size of the traditional test. The author has attempted to estimate the empirical size of the adaptive test for some complex models using 100,000 data sets. Based on these large simulation studies, the true size of the adaptive test appears to be between 4.8% and 5.2% if the errors are from one of the generalized lambda distributions. Thus, the adaptive test appears to maintain its size close enough to the nominal value to be a valid test.

We have shown that the adaptive test is a fair test that can, in some situations, increase the chance of rejecting the null hypothesis when it is false. We now address the practicality of the adaptive method. In the next chapter, software will be described that will allow researchers to easily perform this test using a SAS macro. We will also investigate, with simulation studies, the size and power of the adaptive test when used with certain experimental designs.

Chapter 4

Using Adaptive Tests in the Practice of Statistics

4.1 A SAS Macro for Adaptive Tests

In the previous chapter we described a method of constructing an adaptive test. To make the adaptive test easy to use, a computer program has been developed to perform it, and the SAS macro for this program has been included in Appendix A. In this section we will describe how this macro can be invoked to perform the adaptive test of $H_0 : \beta_A = 0$ versus $H_a : \beta_A \neq 0$ in the linear model

$$Y = X_R\beta_R + X_A\beta_A + \varepsilon_C.$$

Although the SAS macro is lengthy, the user is only required to specify the variable names for the columns in the Y, X_R, and X_A matrices in a one- or two-line statement that will invoke the macro.

An example will illustrate the use of the SAS macro. Suppose we want to perform an adaptive two-sample test using the data from Blumberg, West, and Ellis (1954) given in Chapter 2. This data set contains scores of psychological factors related to the progression of cancer. To perform the adaptive test, we need to use the SAS macro "awls" listed in Appendix A. Before we invoke the macro we need to make a SAS data set that has the independent variable and the dependent variable. For this example we made a permanent SAS data set called "blumberg," which includes the dependent variable "scores" and the independent variable "group." After the macro statements were included in the SAS program and the data set was created, the SAS macro "awls" was invoked by placing the following statement at the end of the program:

```
%awls(dataset=blumberg, y=scores, xr=, nvarsred=0,
      xa=group, classa=group);
```

It is important to include the SAS macro and the SAS statements that make the data set "blumberg" before the macro is invoked. The SAS code in Table 4.1 shows the overall structure of the program that was used to analyze these data. The SAS macro does not execute until it is invoked by the last statement in Table 4.1.

Table 4.1. *The overall layout of a SAS program that contains the "awls" macro.*

```
/*
This file contains a SAS macro for performing the adaptive weighted
least squares test for a subset of a model (macro awls).

****** Macro 1 from Appendix A should be placed here. ******

run;
%mend;

****** Your SAS statements should follow the SAS macro. ******
****** These are used to create the SAS data set. ******

options nocenter linesize=80;
data blumberg;
input scores group;
cards;
-25 1
-24 1

****** Data From Blumberg, West, and Ellis ******

3 2
10 2
;
%awls(dataset=blumberg, y=scores, xr=, nvarsred=0, xa=group, classa=group);
```

In the statement used to invoke the macro the data set is specified by the "dataset" parameter. The dependent variable is specified by the "y" parameter and the variables that are used to make X_R are specified by the "xr" parameter. Since, in this example, the X_R matrix is simply a column of ones, we do not need to specify any variable for the "xr" parameter because the SAS macro will always include an intercept term in the model. The number of variables in the X_R matrix, not including the column of ones that is used for the intercept, is given by the parameter "nvarsred." In this case we have no variables in X_R so we set "nvarsred=0." The "nvarsred" parameter is needed in the weighting procedure to compute the number of degrees of freedom in the t distribution. The variables used in X_A are specified in the "xa" parameter. In this example the variable group is a classification variable, so we would like SAS to make indicator variables corresponding to the levels of that variable. The classification variables in X_A are specified in the "classa" parameter. If we had had classification variables in X_R, we would have indicated these by adding a "classr" parameter. In this invocation of the macro SAS will use a machine-generated "seed" for the random number generator and will, by default, use 2000 permutations. It is not necessary to specify the parameters in the order given in Table 4.1 and in the other examples; the parameters can be specified in any order.

The SAS macro incorporates several options that increase its flexibility. If we want to specify the starting seed for the random number generator, we can use a "seed" parameter. In addition, if we want to specify the number of permutations of the rows of X_A, we can include the "nperm" parameter, as we show in the following example:

```
%awls(dataset=blumberg, y=scores, xr=, nvarsred=0,
      xa=group, classa=group, seed=875456, nperm=5000);
```

The output of this macro is shown in Table 4.2. With 5000 permutations we obtained a p-value of $p = .0670$, which is much smaller than the value we obtained with the traditional analysis.

Table 4.2. *SAS output of macro "awls" for the cancer data.*

```
Adaptive weighted least squares analysis

Number of permutations = 5000 with first seed = 875456
SAS data set: blumberg

Dependent variable: scores

Independent variables in reduced model: Intercept
Class variables in reduced model:

Independent variables in complete model: Intercept group
Class variables in complete model: group

p-value for adaptive test = 0.0670
```

The macro can also be used for more complex models. Suppose we obtain data from human subjects on systolic blood pressure (Y), weight (X_1), age (X_2), and height (X_3). If we want to use systolic blood pressure as the dependent variable we can perform an adaptive test with a multiple regression model

$$Y = \beta_0 + \beta_1 X_1 + \beta_2 X_2 + \beta_3 X_3 + \varepsilon.$$

The macro can be used to test $H_0 : \beta_2 = \beta_3 = 0$ against the alternative H_a: at least one of $\{\beta_2, \beta_3\}$ does not equal zero. If the data have already been placed in the SAS data set "systolic," the macro will be called using the statement

```
%awls(dataset=systolic, y=sbp, xr=weight, nvarsred=1,
      xa=age height, nperm=5000);
```

In this example we used 5000 permutations of the rows of X_A. Because we have not specified the first seed, SAS will obtain it from the computer clock time when the macro is executed. Any number of variables can be specified for the X_R and X_A matrices. Any subsets of the variables in X_R and X_A can be specified as class variables with the "classr" and "classa" variables. There are no limits on the number of permutations, but the execution time will increase with this number. The only required entries are "dataset," "y," "nvarsred," and "xa." An intercept will be included in every model by default, so "xr" does not need to be specified if the reduced model contains only an intercept. The number of observations must not exceed 10,000, but even this could be increased if the SAS macro were modified to permit larger arrays. In the remainder of this chapter we will use several examples to illustrate the use of this "awls" macro.

4.2 An Adaptive Test for Equality of Means in a One-Way Layout

4.2.1 An Example

In this section we will analyze data from a one-way layout using the "awls" macro in Appendix A. Consider the data obtained by Powell and Russell (1984) published by Linton, Edgington, and Davies (1989). These data, which are shown in Table 4.3, represent the dry biomass of ants (in milligrams) measured in the stomach contents of eastern horned lizards, *Phrynosoma douglassi brevirostre*. The data for adult males and yearling females were recorded on four sampling occasions as follows:

Table 4.3. *The dry biomass of ants in the stomach contents of eastern horned lizards.*

Month	Dry Biomass	Median
June	13, 105, 242	105
July	2, 8, 20, 59, 245	20
August	40, 50, 52, 82, 88, 233, 488, 515, 600, 1889	160.5
September	0, 5, 6, 18, 21, 44	12

Suppose we want to use a one-way analysis of variance to determine if the dry biomass varies over the sampling occasions. The traditional way of testing the null hypothesis is to construct the analysis of variance table (Table 4.4). The model for the analysis of variance is

$$y_i = \beta_0 + \beta_1 x_1 + \beta_2 x_2 + \beta_3 x_3 + \varepsilon_i$$

for $i = 1, \ldots, n$, where x_1 is the indicator for June, x_2 is the indicator for July, and x_3 is the indicator for August. In this example we will test the null hypothesis $H_0 : \beta_1 = \beta_2 = \beta_3 = 0$ versus the alternative hypothesis $H_a : \beta_j \neq 0$ for some $j = 1, 2, 3$. Because the p-value associated with the traditional F test is $p = .211$ we have insufficient evidence, with the traditional test, to reject the null hypothesis that the population means are equal.

However, these data, which are shown in the histogram in Figure 4.1, have some very large values in the August measurements that greatly increase the mean squared errors. We

Table 4.4. *An analysis of variance table for the dry biomass of ants data.*

Source	Sum of Squares	d.f.	Mean Square	F
Months	726694.7	3	242231.6	1.64
Error	2947024.2	20	147351.2	
Total	3673719.0	23		

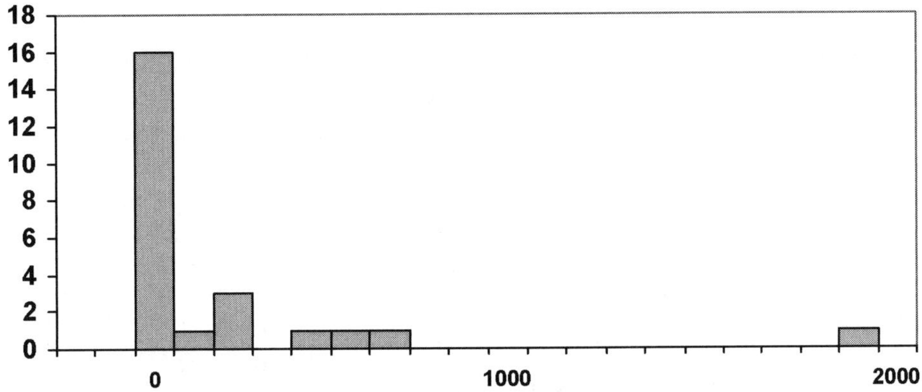

Figure 4.1. *Histogram of the dry biomass of ants data.*

also note that there is very little overlap between the August and September measurements and that the four largest observations in August are much greater than all the observations in the other months. So the raw data suggest that there is some difference between the monthly means, but the traditional F test does not indicate that we should reject the null hypothesis of equality of means.

Manly (1997) analyzed these data using a nonadaptive permutation method and obtained a p-value of $p = .189$ based on $R = 4999$ permutations of the dependent variable. These results show that, for this data set, the traditional F test and the nonadaptive permutation test yield similar results. This is not too surprising since both give full weight to the largest measurement in August, which is more than three times the size of the second-largest observation. We will now use an adaptive test to analyze these data.

4.2.2 Using the SAS Macro

In order to perform the adaptive test of the null hypothesis of equality of means over the months, we used the data on the dry biomass of ants in the stomach of eastern horned lizards to create a SAS data set called "ants," with ant biomass included as the "biomass" variable and the months entered as the character variable "month." We then invoked this SAS macro by including the following statement in the SAS code:

```
%awls(dataset=ants, y=biomass, xr=, nvarsred=0,
      xa=month, classa=month, seed=775, nperm=5000);
```

Since we wanted SAS to make the appropriate indicator variables for the months we used "month" in the "classa=" option. This SAS macro performed the adaptive test using 5000 permutations to produce an adaptive p-value of $p = .0584$. If we had used a significance level of $\alpha = .05$, we would not have rejected the null hypothesis, but we would have some indication that the means may be different. The weights assigned to the observations by the SAS macro are shown in Table 4.5. The variable "Weight" is the weight given to each observation. These weights are used as the diagonal elements of W in the WLS regression. Note that the largest value in August is given a very low weight of $w_{18} = 0.219$ compared to the other observations.

Table 4.5. *The weights assigned to the observations for the ant biomass data.*

Obs.	Biomass	Month	Weight
1	13	June	1.49855
2	242	June	1.35420
3	105	June	1.44499
4	8	July	1.50134
5	59	July	1.47234
6	20	July	1.49463
7	2	July	1.50467
8	245	July	1.35206
9	515	August	1.14880
10	488	August	1.16863
11	88	August	1.45524
12	233	August	1.36060
13	50	August	1.47756
14	600	August	1.08844
15	82	August	1.45882
16	40	August	1.48330
17	52	August	1.47640
18	1889	August	0.21929
19	18	September	1.49575
20	44	September	1.48101
21	21	September	1.49407
22	5	September	1.50301
23	6	September	1.50245
24	0	September	1.50578

For this data set there is a marked difference between the adaptive test results, which are borderline significant, and the traditional F test results, which are clearly not significant. Although the large values of biomass that we observe between August and September may lead us to believe that the adaptive test results are more plausible, it is not prudent to draw a firm conclusion about the effectiveness of the adaptive test from the analysis of one small data set. Consequently, we will now turn our attention to simulation studies in order to fully evaluate the adaptive test when it is used with data from a one-way layout.

4.2.3 Size and Power Studies

In order to determine if the adaptive test could be recommended as a procedure for analyzing data from a one-way layout, we need to compare the size and power of the adaptive test to the size and power of the traditional F test. Let n_j be the sample size in the jth group. To investigate the performance of the adaptive test, a simulation study was performed using a balanced one-way layout having three groups with $n_1 = n_2 = n_3 = 8$, a balanced one-way layout having four groups with $n_1 = n_2 = n_3 = n_4 = 15$, and an unbalanced one-way layout having four groups with $n_1 = 30$, $n_2 = n_3 = n_4 = 10$. In these studies we used 10,000 data sets for each configuration of sample size and error distribution. The error distributions are the same nine error distributions used in Chapter 3 for the two-sample test. For each data set generated, we used $R = 2000$ permutations in the adaptive test. We show in Table 4.6 the empirical size for the adaptive test and for the traditional F test when $\alpha = .05$ was used for both tests. Note that both tests had empirical sizes between 4.3% and 5.4% over the sample size configurations and error distributions. Thus, it appears that the adaptive test maintains its size near the nominal significance level for the one-way layout.

Table 4.6. *Empirical size in percent for the adaptive test and the F test for equality of means in the one-way layout.*

	Distributions								
Skewness	0	0	0	1	1	1	2	2	2
Kurtosis	1.8	3.0	6.0	4.2	5.4	8.4	11.4	12.6	15.6
	$n_1 = n_2 = n_3 = 8$								
Adaptive	4.7	5.0	4.8	4.9	4.8	5.4	4.9	5.0	5.1
F test	5.0	5.0	4.8	4.5	4.6	4.9	4.1	4.0	4.5
	$n_1 = n_2 = n_3 = n_4 = 15$								
Adaptive	4.8	5.3	5.1	4.6	5.0	4.9	5.3	5.3	5.3
F test	4.9	5.2	4.9	4.6	4.7	4.6	4.5	4.6	4.8
	$n_1 = 30,\ n_2 = n_3 = n_4 = 10$								
Adaptive	4.7	5.1	4.7	5.0	4.6	5.1	5.3	5.0	4.3
F test	4.8	5.3	4.6	5.0	4.8	4.9	4.9	5.0	4.4

Table 4.7 gives the empirical power for the adaptive test and the F test for the same study sizes and error distributions that were used to produce the results in Table 4.6. The simulations were designed so that the population mean in the first group was shifted to the left of the means of the other groups. Except for the first group, the means were equal. The shift parameter was determined so that the power of the F test would approach 50% for normal errors. In these simulations we used $R = 2000$ permutations for each of the 10,000 data sets. These results show that the adaptive test is more powerful than the traditional F test for nonnormal errors. Even with $n_1 = n_2 = n_3 = 8$ it appears that the adaptive test is the best choice unless the researcher is quite certain that the error distribution is normal or approximates the normal. It appears that the adaptive test can be recommended for routine use with the one-way layout provided the sample size equals or exceeds 8 in each group.

Table 4.7. *Empirical power in percent for the adaptive test and the F test for equality of means in a one-way layout.*

	Distributions								
Skewness	0	0	0	1	1	1	2	2	2
Kurtosis	1.8	3.0	6.0	4.2	5.4	8.4	11.4	12.6	15.6
	$n_1 = n_2 = n_3 = 8$								
Adaptive	49.8	49.2	55.0	57.4	56.8	58.7	68.2	67.3	65.3
F test	48.5	50.1	52.3	52.0	52.9	54.6	56.6	57.0	56.1
	$n_1 = n_2 = n_3 = n_4 = 15$								
Adaptive	57.1	48.8	56.2	63.7	60.7	61.8	77.0	75.0	71.9
F test	48.7	49.7	51.8	51.0	51.7	53.1	53.9	54.9	55.8
	$n_1 = 30,\ n_2 = n_3 = n_4 = 10$								
Adaptive	58.3	49.6	56.3	61.5	59.4	60.6	75.3	72.2	69.8
F test	48.9	50.2	51.8	50.3	51.0	52.3	52.5	52.5	52.7

If the total sample size approximates 60 and the error distributions are nonnormal, then the adaptive test has a considerable power advantage over the traditional F test.

4.3 An Adaptive Test for the Slope in a Simple Linear Regression

4.3.1 An Example

In this section we show how the adaptive test can be used to perform a test for slope in a simple linear regression. That is, in the model

$$y_i = \beta_0 + \beta_1 x_i + \varepsilon_i$$

we will use adaptive methods to test $H_0 : \beta_1 = 0$ versus $H_1 : \beta_1 \neq 0$. The traditional test is based on the p-value obtained from a t statistic. However, the traditional test, which is the most powerful test when the errors are normally distributed, may not be the most powerful test otherwise.

The following example will illustrate the value of the adaptive test for the slope in a simple linear regression. The ACT assessment test is designed to measure the ability of high school students to perform college work and to determine their general educational development. The test is given to many high school seniors in the United States. Suppose an educator is interested in comparing the average ACT scores in a state to the average scores in other states. One of the main problems in the interpretation of these data is that the average in a state may depend, in part, on the percentage of students who took the ACT in that state. One theory is that the average ACT scores will be inflated in those states with a small percentage of students who take the ACT test, rather than the competing test, because most of those taking the ACT test will be more highly motivated students who are considering a variety of colleges.

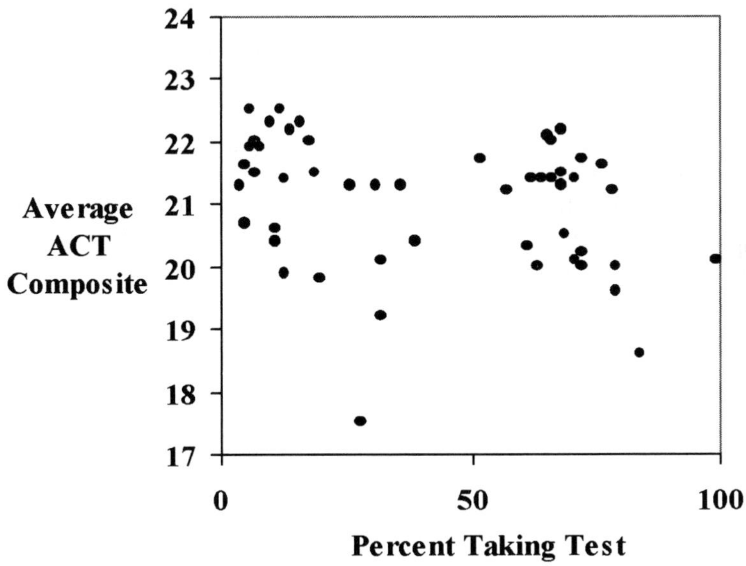

Figure 4.2. *Scatterplot of average ACT composite score and percent taking test.*

To test this theory we have obtained the average ACT composite scores and the percent of graduates tested for the year 2002 for 50 states and the District of Columbia in the United States (ACT, 2002). The scatterplot of the average ACT composite scores and the percent of graduates tested for the year 2002 is shown in Figure 4.2.

The scatterplot does not indicate a clear relationship between the percent of test takers and the average ACT composite score, but it does appear that states with a higher percentage of test takers tend to have lower scores than states with a small percentage of test takers. The point with the lowest ACT score represents the District of Columbia, and the second-lowest score is from Mississippi.

4.3.2 Using the SAS Macro

The traditional t test for slope yields a p-value of $p = .0356$, which suggests that the average ACT scores decline with increased percentages of individuals taking the test. However, because there are some unusual values in the data set, we would like to see if the adaptive test will produce a clearer result. To perform the adaptive test we use the SAS macro

```
%awls(dataset=act, y=score, xr=, xa=percent,
      seed=2333, nperm=2000);
```

In this invocation of the SAS macro "act" is the name of the SAS data set, "score" is the variable containing the average ACT composite score, and "percent" is the percentage of test takers in the state. The adaptive test for the slope gives a p-value of $p = .0115$, which indicates more clearly than the traditional test that there is a negative correlation between the average ACT score and the percent of takers. In the adaptive test the observation from the District of Columbia, which was the most extreme outlier, was given the smallest weight of $w_8 = 0.608$, while the other observations received weights in the interval [0.753, 1.229].

Some researchers may attempt to remove a few outliers from this data set before performing a test for slope, but it is not always clear what effect the removal of outliers might have on the significance level of a test after the adjustments are made to the data. The advantage of the adaptive test is that it downweights, without human intervention, observations that are judged to be outliers in a manner that ensures that the resulting test will maintain its level of significance.

4.3.3 Size and Power Studies

In order to document the performance of the adaptive test for slope, simulation studies were used to compare its size and power to the t test for slope. In these simulations the x-values were generated from a normal distribution. For specified values of β_0 and β_1 the mean of the ith observation equaled $\beta_0 + \beta_1 x_i$, and the y-value was obtained by generating a random error ε_i from a generalized lambda distribution and adding it to the mean value to obtain $y_i = \beta_0 + \beta_1 x_i + \varepsilon_i$ for $i = 1, \ldots, n$. In these simulations we used 10,000 data sets and $R = 2000$ permutations.

For the simulations used to determine the size of the test we set $\beta_1 = 0$. The traditional test is the t test for $H_0 : \beta_1 = 0$ against $H_a : \beta_1 \neq 0$ in a simple linear regression, which will give the same p-value as that obtained from an F test. The empirical size of the adaptive test is given in Table 4.8 and ranged from 4.6% to 5.3% over the three sample sizes and nine distributions of error terms. Thus, it appears that the adaptive test maintains its size.

Table 4.8. *Empirical size estimates in percent for the adaptive test and the t test for slope in a simple linear regression.*

	Distributions								
Skewness	0	0	0	1	1	1	2	2	2
Kurtosis	1.8	3.0	6.0	4.2	5.4	8.4	11.4	12.6	15.6
					$n = 10$				
Adaptive	4.6	5.1	5.1	5.1	4.8	5.1	4.6	5.0	5.0
t test	4.6	5.0	5.0	5.2	5.0	5.1	4.7	4.9	5.0
					$n = 20$				
Adaptive	5.3	5.3	5.0	5.0	5.1	5.2	5.2	4.8	5.1
t test	5.0	5.0	5.0	5.2	5.1	5.1	4.9	4.7	5.1
					$n = 50$				
Adaptive	5.1	5.2	5.1	4.9	5.0	4.8	4.7	4.9	5.2
t test	5.1	5.0	5.1	4.8	5.4	4.9	5.1	4.8	5.1

For the power estimates the slope was set to $\beta_1 = 2/\sqrt{n}$ so that the power of the traditional test would not be too far from 50%. The empirical power estimates are shown in Table 4.9 for the same sample sizes and error distributions as were used in the simulations for the size of the test. Again, the x-values were generated using the approximate normal

Table 4.9. *Empirical power estimates in percent for the adaptive test and the t test for slope in a simple linear regression model.*

	Distributions								
Skewness	0	0	0	1	1	1	2	2	2
Kurtosis	1.8	3.0	6.0	4.2	5.4	8.4	11.4	12.6	15.6
					$n = 10$				
Adaptive	34.0	35.6	42.0	41.3	41.4	44.8	50.2	49.4	48.3
t test	34.7	37.5	42.1	40.1	41.3	43.9	46.6	46.6	45.6
					$n = 20$				
Adaptive	44.2	43.0	49.0	51.0	51.5	52.6	62.2	60.1	60.0
t test	43.3	44.5	46.8	46.5	47.8	48.9	50.5	50.4	52.5
					$n = 50$				
Adaptive	56.9	47.7	53.9	61.0	57.7	59.2	73.5	71.0	68.3
t test	48.5	49.2	49.9	49.5	50.2	51.7	52.8	52.4	53.4

distribution when $R = 2000$ permutations were used with 10,000 data sets. For the samples with $n = 10$ observations there was little difference in power between the adaptive test and the traditional t test. However, for samples of size $n = 50$ the adaptive test had a considerable power advantage for all nonnormal distributions. Consequently, it appears that the adaptive test for slope is recommended if the sample size equals or exceeds 20.

4.4 An Adaptive Test for Regression Coefficients in a Multiple Regression

4.4.1 An Example

Are adaptive tests effective when used with complex models? To investigate this we will test the null hypothesis that two regression coefficients equal zero in a multiple regression model with four independent variables. For our example we will use the cigarette consumption data for the year 1970 listed in Appendix B for 50 states and the District of Columbia. These data were analyzed by Chatterjee, Hadi, and Price (2000). The dependent variable is the per capita cigarette sales (in packs). The independent variables are the average price for a pack, the per capita personal income, the median age, and the percent of females living in the state.

Suppose that preliminary analysis indicates that the price and income are predictive of cigarette sales. Further, suppose we want to determine if age and gender are related to sales after controlling for price and income. The complete model is then

$$Y = \beta_0 + \beta_1 X_1 + \beta_2 X_2 + \beta_3 X_3 + \beta_4 X_4 + \varepsilon,$$

where Y = sales, X_1 = price, X_2 = income, X_3 = age, and X_4 = percent female. The p-values for the individual coefficients on the SAS output for the complete model, which

Table 4.10. *SAS output from PROC GLM for the complete model.*

Source	DF	Sum of Squares	Mean Square	*F* Value	Pr > *F*
Model	4	15765.67788	3941.41947	5.08	0.0018
Error	46	35659.76565	775.21230		
Corrected Total	50	51425.44353			
Source	DF	Type I SS	Mean Square	*F* Value	Pr > *F*
Price	1	4647.521381	4647.521381	6.00	0.0182
Income	1	8223.188364	8223.188364	10.61	0.0021
Age	1	2723.715504	2723.715504	3.51	0.0672
Gender	1	171.252634	171.252634	0.22	0.6406

is given in Table 4.10, suggest that age and gender may not be needed in the model. To determine if age and gender are useful we test $H_0 : \beta_3 = \beta_4 = 0$ against $H_a : \beta_3 \neq 0$ or $\beta_4 \neq 0$ by using the traditional F test for a subset of regression coefficients. For the complete model we obtain $SSE_C = 35831.0$, and for the reduced model $Y = \beta_0 + \beta_1 x_1 + \beta_2 x_2 + \varepsilon$ we obtain $SSE_R = 38554.7$. Consequently, if we use the traditional F test for a subset of regression coefficients, we obtain a test statistic of $F = 1.87$, which yields a p-value of $p = .1661$. Thus, with the traditional F test there is no reason to reject the null hypothesis.

4.4.2 Using the SAS Macro

In order to perform an adaptive test of $H_0 : \beta_3 = \beta_4 = 0$ with the model

$$Y = \beta_0 + \beta_1 X_1 + \beta_2 X_2 + \beta_3 X_3 + \beta_4 X_4 + \varepsilon,$$

we will use the SAS macro. If the data set is named "cigs" and we want to perform 5000 permutations of the rows of X_A, we can invoke the macro with the following SAS code:

```
%awls(dataset=cigs, y=sales, xr=price income, nvarsred=2,
      xa=age gender, seed=3433, nperm=5000);
```

The output of the SAS macro shown in Table 4.11 indicates that the p-value for the adaptive test is $p = .0264$. The adaptive test clearly indicates that at least one of the candidate variables in X_A should be included in the model.

Table 4.11. *Output from the SAS macro for the adaptive test of* $H_0 : \beta_3 = \beta_4 = 0$.

```
Adaptive weighted least squares analysis

Number of permutations = 5000 with first seed = 3433
SAS data set: cigs

Dependent variable: sales

Independent variables in reduced model: Intercept price income
Class variables in reduced model:

Independent variables in complete model: Intercept price income age
                                          gender
Class variables in complete model:

p-value for adaptive test = 0.0264
```

The dramatic difference in the p-values between the adaptive test and the traditional test is due primarily to the unusually large sales of cigarettes in New Hampshire and Nevada. The adaptive test gave a weight of $w_{30} = 0.363$ to New Hampshire and $w_{29} = 0.694$ to Nevada, while all other weights were in the interval [1.03, 1.50]. Because the adaptive test indicated that either age or gender is related to sales, after controlling for price and income, we used an adaptive test of $H_0 : \beta_3 = 0$ in the model that included all four independent variables and obtained a p-value of $p = .2109$. The adaptive test of $H_0 : \beta_4 = 0$ in the four independent variable model gave a p-value of $p = .1624$. Based on these p-values we proceeded to remove age from the model.

We then turned our attention to testing $H_0 : \beta_3 = 0$ in the three-variable model

$$Y = \beta_0 + \beta_1 X_1 + \beta_2 X_2 + \beta_3 X_3 + \varepsilon,$$

where Y = sales, X_1 = price, X_2 = income, and X_3 = gender. The traditional F test gave a p-value of $p = .1433$, whereas the adaptive test gave a p-value of $p = .0190$. Consequently, based on the results from the adaptive test, we concluded that gender should be included in the model. This example illustrates how the adaptive test can be used for a variable selection procedure.

Although the adaptive approach appears to be helpful with the analysis of these data, there are several difficulties that need to be addressed. In the adaptive test of $H_0 : \beta_3 = \beta_4 = 0$ in the four-variable model we permute the rows of X_A, which destroys the relationship between the independent variables in X_A and those in X_R. Will this permutation method maintain its level of significance when X_A has two columns? Also, will the adaptive test be reasonably powerful when there are correlations between the independent variables? We will answer these questions by evaluating the performance of the adaptive test in multiple regression models when X_A has two columns.

4.4.3 Size and Power Studies

A simulation study was performed to determine the effectiveness of the adaptive test when several variables are considered as candidate variables in a multiple regression model. In these simulations four independent variables $\{X_1, X_2, X_3, X_4\}$ were generated from lognormal distributions for use in the model $y = \beta_0 + \beta_1 x_1 + \beta_2 x_2 + \beta_3 x_3 + \beta_4 x_4 + \varepsilon$. The variables $\{X_1, X_3\}$ were generated from a bivariate lognormal distribution with a correlation of $\rho = 0.8$. The variables $\{X_2, X_4\}$ were also generated from a bivariate lognormal distribution with a correlation of $\rho = 0.8$. By generating the variables in this way we introduced a large correlation between the variables in X_R and the variables in X_A. We then determined, via simulation techniques, the empirical size and power of the adaptive test of the null hypothesis $H_0 : \beta_3 = \beta_4 = 0$ against the alternative $H_a : \beta_3 \neq 0$ or $\beta_4 \neq 0$.

The errors were generated from the same nine members of the generalized lambda distribution that were used for the other simulations. The empirical size is listed in Table 4.12 for the adaptive test and the traditional F test for $n = 20$ and $n = 50$ observations. These results show that the adaptive test has a significance level close to the nominal value of $\alpha = .05$, which is encouraging if we consider the high correlation between the independent variables. We also note that the traditional F test often has an empirical size that exceeds the nominal value when the errors are skewed.

Table 4.12. *Empirical size in percent of the adaptive test and the F test for testing the null hypothesis $H_0 : \beta_3 = \beta_4 = 0$ in a multiple linear regression model. The independent variables were generated from bivariate lognormal distributions with a correlation of 0.8 between X_1 and X_3 and between X_2 and X_4.*

	Distributions of the Errors								
Skewness	0	0	0	1	1	1	2	2	2
Kurtosis	1.8	3.0	6.0	4.2	5.4	8.4	11.4	12.6	15.6
					$n = 20$				
Adaptive	4.7	5.0	5.2	4.8	5.1	4.8	4.8	5.2	5.4
F test	4.3	5.2	5.5	5.2	5.4	5.5	5.5	6.1	6.3
					$n = 50$				
Adaptive	5.3	4.8	5.0	4.8	5.1	4.8	5.0	5.2	5.3
F test	4.6	4.9	5.3	5.0	5.6	5.5	5.8	5.8	6.0

The empirical power of the adaptive test is given in Table 4.13 for the same sample sizes and distributions as were used for the size estimates. The slope was set to $4/\sqrt{n}$ for each independent variable in these power simulations so that the power of the F test would be between 50% and 75%. These results show that there is an advantage to using the adaptive test if $n = 50$ and the errors are nonnormal. In addition, if $n = 20$, the adaptive test is slightly more powerful than the F test when the errors are skewed, while it has an empirical size that is closer to $\alpha = .05$. Thus, for $n \geq 20$ the adaptive test is recommended, unless the researcher knows that the errors are normally distributed.

Table 4.13. *Empirical power in percent of the adaptive test and the F test for testing the null hypothesis $H_0 : \beta_3 = \beta_4 = 0$ in a multiple linear regression model. The independent variables were generated from bivariate lognormal distributions with a correlation of 0.8 between X_1 and X_3 and between X_2 and X_4.*

	Distributions of the Errors								
Skewness	0	0	0	1	1	1	2	2	2
Kurtosis	1.8	3.0	6.0	4.2	5.4	8.4	11.4	12.6	15.6
					$n = 20$				
Adaptive	50.3	51.0	55.5	54.0	55.0	56.7	62.5	61.2	60.2
F test	51.6	53.8	55.7	53.2	54.7	56.1	57.7	57.2	57.1
					$n = 50$				
Adaptive	73.2	67.6	73.1	75.5	74.9	75.8	85.1	83.4	82.8
F test	69.5	69.4	70.7	69.4	70.5	70.5	71.2	70.5	71.0

4.5 An Adaptive Test for Interaction with a Two-Way Layout

4.5.1 An Example

In section 4.2 we considered the adaptive analysis of data from a one-way layout. In this section we consider the more complex models required to analyze data from a two-way layout. Consider, for example, data on the number of insects collected by a square-foot bottom sampler in Shope Creek and Ball Creek, North Carolina. These data, which are shown in Table 4.14, were collected in December 1952 and March, June, and September of 1955 by W. Hassler and were used as an example in Simpson, Roe, and Lewontin (1960, p. 284). The main effects are creek and month. In this section we want to use adaptive methods to see if there is any interaction between the creek and the month. The traditional F test for interaction produces a p-value of $p = .0467$, which is borderline significant. This suggests that the differences that we observe between the two creeks may vary over time. We will use the adaptive test in an effort to more clearly determine if the interaction is significant.

4.5.2 Using the SAS Macro

The SAS macro can be used to perform an adaptive test for interaction in a two-way layout. However, the user must set up appropriate indicator variables for the interaction effect to be used in the macro call. If we use a traditional test, we can let SAS create, in PROC GLM, the appropriate indicator variables for the interaction effect. Unfortunately, PROC GLM does not actually create the interaction terms but only computes the appropriate elements in the $(X'X)$ matrix. Thus, if we perform an adaptive test, the adaptive macro uses PROC GLM, which does not create the indicator variables for the interaction terms. So, if we use

Table 4.14. *The number of insects in a square-foot sampler by creek and month. Six replicates were obtained at both creeks in each month.*

Obs.	Month	Creek	Insects	Obs.	Month	Creek	Insects
1	December	Shope	7	25	June	Shope	124
2	December	Shope	9	26	June	Shope	51
3	December	Shope	19	27	June	Shope	63
4	December	Shope	1	28	June	Shope	81
5	December	Shope	18	29	June	Shope	83
6	December	Shope	15	30	June	Shope	106
7	December	Ball	25	31	June	Ball	20
8	December	Ball	9	32	June	Ball	44
9	December	Ball	16	33	June	Ball	26
10	December	Ball	28	34	June	Ball	127
11	December	Ball	10	35	June	Ball	38
12	December	Ball	14	36	June	Ball	52
13	March	Shope	29	37	September	Shope	72
14	March	Shope	37	38	September	Shope	87
15	March	Shope	114	39	September	Shope	100
16	March	Shope	49	40	September	Shope	68
17	March	Shope	24	41	September	Shope	67
18	March	Shope	64	42	September	Shope	9
19	March	Ball	35	43	September	Ball	40
20	March	Ball	45	44	September	Ball	45
21	March	Ball	22	45	September	Ball	263
22	March	Ball	29	46	September	Ball	100
23	March	Ball	18	47	September	Ball	129
24	March	Ball	27	48	September	Ball	115

an adaptive test for interaction, which requires the permutation of interaction effects, we must code our own interaction effects.

In our example we do not need to set up indicator variables for the main effects of month and creek, since SAS will create the appropriate indicator variables if we specify them as classification variables in the macro call. However, we must set up our own interaction terms. One way of coding our interaction terms is as follows:

$$IND_1 = \begin{cases} 1 & \text{if Month = March and Creek = Ball,} \\ 0 & \text{otherwise;} \end{cases}$$

$$IND_2 = \begin{cases} 1 & \text{if Month = June and Creek = Ball,} \\ 0 & \text{otherwise;} \end{cases}$$

$$IND_3 = \begin{cases} 1 & \text{if Month = September and Creek = Ball,} \\ 0 & \text{otherwise.} \end{cases}$$

In order to perform the adaptive test, we use the main effects of month and creek in the reduced model. For the complete model we add the indicator variables for the interaction effect. Since there are four levels for the month effect and two levels for the creek effect, the usual parameterization will include four variables for the main effects, not including the intercept. So we will set "nvarsred=4" and the macro call will be

```
%awls(dataset=creek, y=insects, xr=month creek,
      classr=month creek, nvarsred=4, xa = ind1 ind2 ind3,
      seed = 31739, nperm = 2000);
```

The weights shown in Figure 4.3 indicate that the third replicate taken at Ball Creek in the month of September will be downweighted for the adaptive analysis. The adaptive p-value is $p = .0315$ for the interaction effect, which indicates more clearly than the traditional test that there may be some interaction between month and creek.

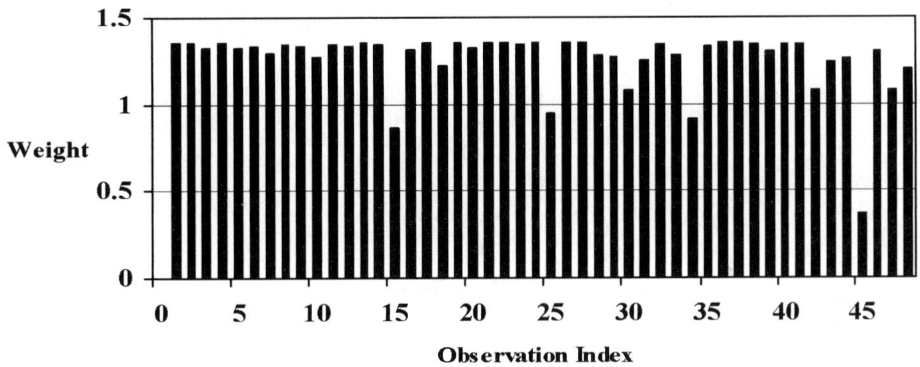

Figure 4.3. *Weights for the adaptive test for the interaction effect.*

The result from the adaptive test is encouraging, but we need to carefully evaluate its performance. For a test of interaction the permutation method may appear to be inappropriate because, after permuting the rows in X_A, the rows in the permuted matrix will not be consistent with the main effects that were coded in X_R. In our example, the interaction effects are coded only for Ball Creek, but they may, after permutation, be found with an observation from Shope Creek. Thus, we need to use a simulation study to investigate the size and power of these tests for interaction in several two-way layouts.

4.5.3 Size and Power Studies

A simulation study was performed to evaluate the performance of the adaptive test for interaction. In these simulations we used a 2×2 layout and let n_{ij} = number of observations corresponding to the ith level of the first factor and the jth level of the second factor. Table 4.15 gives the empirical size and Table 4.16 gives the empirical power in percent for the test for interaction in a two-way design with two balanced designs and one unbalanced

Table 4.15. *Empirical size in percent for the adaptive test and the F test for the interaction terms in a two-way design. In these designs each factor had two levels. The cell count for the ith row and jth column is indicated by n_{ij}.*

	Distributions of the Errors								
Skewness	0	0	0	1	1	1	2	2	2
Kurtosis	1.8	3.0	6.0	4.2	5.4	8.4	11.4	12.6	15.6
	Balanced design with $n_{11} = n_{12} = n_{21} = n_{22} = 8$								
Adaptive	5.1	5.2	4.9	4.9	4.8	5.1	4.9	4.5	4.8
F test	4.8	5.3	4.8	4.9	4.8	5.0	4.5	4.3	4.8
	Balanced design with $n_{11} = n_{12} = n_{21} = n_{22} = 16$								
Adaptive	5.2	4.6	4.9	4.7	5.5	5.3	4.4	4.9	4.8
F test	5.1	4.4	4.9	4.9	5.4	5.2	4.2	4.7	4.9
	Unbalanced design with $n_{11} = 40,\ n_{12} = n_{21} = n_{22} = 8$								
Adaptive	5.5	4.4	5.4	5.1	4.9	4.9	4.8	4.9	5.0
F test	5.3	4.4	5.3	5.2	4.9	4.9	5.3	5.0	4.8

Table 4.16. *Empirical power in percent for the adaptive test and the F test for the interaction terms in a two-way design. In these designs each factor had two levels. The cell count for the ith row and jth column is indicated by n_{ij}.*

	Distributions of the Errors								
Skewness	0	0	0	1	1	1	2	2	2
Kurtosis	1.8	3.0	6.0	4.2	5.4	8.4	11.4	12.6	15.6
	Balanced design with $n_{11} = n_{12} = n_{21} = n_{22} = 8$								
Adaptive	50.6	47.1	54.3	56.8	55.5	57.3	68.4	66.0	63.5
F test	47.7	48.5	51.6	50.1	50.7	52.5	53.9	54.0	53.1
	Balanced design with $n_{11} = n_{12} = n_{21} = n_{22} = 16$								
Adaptive	58.5	48.9	56.0	62.7	59.2	60.9	75.6	73.0	68.9
F test	49.2	50.2	51.6	51.4	51.4	53.4	54.1	52.9	53.1
	Unbalanced design with $n_{11} = 40,\ n_{12} = n_{21} = n_{22} = 8$								
Adaptive	67.0	59.1	64.1	71.2	68.1	67.7	83.0	80.6	78.2
F test	58.5	60.3	60.4	60.3	61.2	60.9	63.1	62.8	63.8

design. For the simulations shown in Table 4.15 the sizes ranged from 4.4% to 5.5%, which demonstrates that both the adaptive test and the traditional F test maintain the $\alpha = .05$ level of significance.

Because we suspected that the size of the adaptive test for interaction might depart somewhat from the nominal value, we performed a large simulation study using 100,000 data sets to accurately estimate the size with $n = 64$ observations in the unbalanced design specified in the last part of Table 4.15. A simulation size of 100,000 data sets gave a standard error of 0.07%. Based on this large simulation the empirical size of the adaptive test was 4.86% when the error terms were generated from the distribution with $\alpha_3 = 1$ and $\alpha_4 = 5.4$.

Another large simulation using 100,000 data sets showed that the empirical size of the test was 5.02% when the error terms were generated from the distribution with $\alpha_3 = 2$ and $\alpha_4 = 12.6$. These results do not show that the adaptive test for interaction maintains the significance level exactly, but they do show that it has an empirical size that is very close to the nominal value with unbalanced data.

The empirical power estimates given in Table 4.16 demonstrate that the adaptive test had greater power than the traditional test for models with nonnormal errors. These results for the adaptive and traditional tests are similar to those seen in the other simulation studies comparing these two tests. Also, as we have seen with other simulation studies, the F test has only a small power advantage over the adaptive test with normal errors.

4.6 An Adaptive Test for the Treatment Effect in a Multicenter Clinical Trial

As part of the drug approval process, pharmaceutical companies sometimes use multicenter clinical trials to evaluate the safety and efficacy of a promising drug. Several centers are used because a single center may have difficulty enrolling a sufficient number of subjects. In this section we present an adaptive test for the treatment effect in a stratified design, such as a multicenter clinical trial, when the outcome measure is continuous.

In these clinical trials the center effect is really not of great interest. Usually, the purpose of the clinical trial is to determine the effectiveness of a treatment. If there is a potential interaction between the center effect and the treatment effect, then the adaptive test for interaction described in section 4.5 could be used. In this section we will assume that we have already determined that an interaction term is not appropriate for the model so we will limit our inference to the common treatment effect. Further, we will consider the centers to be fixed effects, since in many study protocols the centers are not selected from a larger population. For a discussion of many issues involved in the analysis of continuous and discrete data from clinical trials see Fleiss (1986, Chap. 6) and Davis and Chung (1995).

Suppose that individuals are randomly assigned to one of two treatments in each of c centers. Let y_{ijk} be the continuous outcome measure for the kth individual who was given the jth treatment at the ith center. The linear model that we will use is

$$y_{ijk} = \beta_0 + \beta_1 x_{1,jk} + \cdots + \beta_{c-1} x_{c-1,jk} + \beta_c x_{ijk} + \varepsilon_{ijk},$$

where ε_{ijk} is the error term, the set $\{x_{1,jk}, \ldots, x_{c-1,jk}\}$ contains indicator variables for the first $c - 1$ centers, and x_{ijk} is the treatment indicator. In this model the treatment effect is represented by β_c. The test of significance for the treatment effect is usually performed by testing $H_0 : \beta_c = 0$ against $H_a : \beta_c \neq 0$ with an F test for a subset of coefficients in a linear model.

4.6.1 An Example

The following hypothetical data will illustrate how the adaptive test could be used to analyze data from a multicenter clinical trial. In order to investigate the effectiveness of an appetite suppressant in a diet and exercise program, suppose that approximately 20 individuals were randomly assigned to either a drug group or a placebo group in each of the 6 centers.

Subjects who enrolled in this study were also given general nutritional advice and exercise recommendations. At the end of 12 weeks the weight loss (in kilograms) was recorded for each individual. These hypothetical weight loss measurements are recorded in Table 4.17 along with the mean values for each center-treatment combination. A positive value indicates a loss in weight; a negative value indicates a weight gain. Note that few individuals lost more than 6 kg of body weight under the placebo or the drug treatment.

Table 4.17. *Weight loss (in kilograms) for individuals assigned to either a placebo group or a drug group in a multicenter clinical trial.*

Center	Placebo Weight Loss	Placebo Mean	Drug Weight Loss	Drug Mean
1	2.4, 2.7, −1.5, 3.7, 2.7 2.3, 3.4, 3.6, 4.9 4.7	2.89	2.2, 4.7, 2.9, 2.8, 5.3 4.1, 4.5, 2.4, 2.0	3.43
2	3.9, 4.4, 4.3, 4.9, 4.0 13.3, 2.4, 1.8, 4.0	4.78	5.7, 5.7, −0.7, 4.2, 4.1 5.1, 2.7, 4.9, 4.7, 5.5 3.5, 4.5	4.16
3	2.1, 1.2, 2.9, 4.6, 1.6 10.5, 1.0, 2.6, −6.2, 2.7	2.3	4.1, 3.8, 5.5, 5.6, 4.4 2.6, 5.6, 5.0, 4.7	4.59
4	2.9, 1.3, 4.9, 2.5, 2.4 3.7, 1.2, 2.8, 1.2, 2.6	2.55	1.9, 5.0, 2.4, 1.9, −5.6 2.0, 4.7, 4.2, 2.9, 4.6	2.40
5	4.0, 3.1, 4.6, 4.5, 4.9 4.5, 1.8, 1.3, 3.6	3.58	3.8, 2.6, 4.2, 5.3, 5.0 14.7, 5.2, 5.7, 2.7, 5.6	5.48
6	1.8, 2.3, 4.4, 2.2, 1.1 4.0, −1.3, 3.7, 4.8, 4.2	2.72	4.7, 2.0, 4.2, −0.6, 5.0 4.6, 5.0, 4.3, 2.9	3.57

The data do not clearly indicate that the appetite suppressant was effective, since the weight loss with the drug exceeded the weight loss with the placebo in only four of the six centers. In addition, the traditional F test, which was computed using PROC GLM in SAS with center and treatment as main effects, gave a p-value of $p = .0845$. This p-value would not allow us to reject the null hypothesis if we used $\alpha = .05$. We note that most subjects lost between 1 kg and 5 kg, but three individuals lost more than 10 kg and two individuals gained more than 5 kg. We will now use the adaptive test to see if the presence of these five outliers had a great influence on the test results.

4.6.2 Using the SAS Macro

To perform the adaptive test we will use the "awls" macro. Suppose we create a SAS data set called "suppress," a weight loss variable called "wloss," a center variable called "center," and a drug treatment variable called "treatmnt." If we give the "center" variable the center number, we can generate indicator variables for the center by using "center" as a classification variable in the macro call. We do this by using "center" in the "xr=" and "classr=" options. With six centers we have five indicator variables in the reduced model, not counting the intercept, so we set "nvarsred" to five. The treatment effect will also be used as a classification variable by coding it in the "xa=" and "classa=" options. The SAS macro can be invoked with the following macro call:

```
%awls(dataset=suppress, y=wloss, xr=center,
      classr=center, nvarsred=5, xa=treatmnt,
      classa=treatmnt, seed=1731, nperm=2000);
```

The p-value for the adaptive test was computed to be $p = .003$, which indicates that the appetite suppressant was an effective weight loss medication. Most of the weights for the observations were in the interval [1.25, 1.45]. The observations of 13.3 in the second center and 14.7 in the fifth center were both given a weight of 0.57. In addition, the individuals who gained more than 5 kg were downweighted because these observations were given low observational weights in the interval [0.6, 0.7]. The adaptive method produced a small p-value because it automatically downweighted the extreme observations.

However, we cannot tell, from the analysis of one data set, if the adaptive method is really effective. In our example we have approximately ten observations for each treatment in each center. Will this method work when the number of centers is large and the number of individuals in each center is small? We will now use a simulation study to determine the effectiveness of the adaptive test with several study configurations.

4.6.3 Size and Power Studies

In these Monte Carlo simulations we used $c = 5$ centers with $n = 80$ total patients and $c = 20$ centers with $n = 80$ and $n = 400$ total patients. Each center had (n/c) patients, and these patients were assigned at random to one of the two treatments. Table 4.18 gives the empirical size of the adaptive test and the F test for three combinations of n and c and for the nine simulations corresponding to the nine distributions of error terms. As we have done with other simulation studies, we used $\alpha = .05$ with 10,000 data sets to estimate the size and power of the tests. The results in Table 4.18 indicate that the adaptive test and the F test maintained their level of significance.

The empirical power in percent for the adaptive test and the F test is shown in Table 4.19. For these simulations we set $\beta_c = 3.29/\sqrt{n}$, which gave a power near 40% for the F test. For nonnormal errors the adaptive test was clearly more powerful than the traditional F test with $n = 80$ and $c = 5$, which had eight observations for each treatment in each center. The adaptive test was also more powerful with $n = 400$ and $c = 20$, which had ten observations for each treatment in each center. As we have seen with other models, the F test has only a small power advantage over the adaptive test when the errors are normal. However, for $n = 80$ and $c = 20$, which had only two observations for each treatment in each center, the adaptive test was only slightly more powerful than the F test. We conclude that the adaptive test is recommended if there are at least 80 patients in the study.

Table 4.18. *Empirical size in percent for the adaptive test and the F test for the treatment effect in a multicenter clinical trial. The total number of observations is indicated by n and the number of centers is indicated by c.*

	Distributions of the Errors								
Skewness	0	0	0	1	1	1	2	2	2
Kurtosis	1.8	3.0	6.0	4.2	5.4	8.4	11.4	12.6	15.6
	$n = 80,\ c = 5$								
Adaptive	5.3	5.3	4.9	5.0	4.9	5.4	5.2	4.7	5.1
F test	5.1	5.2	4.8	4.8	4.8	5.0	4.9	4.6	4.8
	$n = 80,\ c = 20$								
Adaptive	5.4	5.2	5.3	4.5	5.3	5.3	4.9	5.0	5.0
F test	5.4	5.3	5.2	4.4	5.1	5.3	4.3	4.7	4.7
	$n = 400,\ c = 20$								
Adaptive	5.2	5.3	5.2	4.6	5.3	4.5	5.5	5.1	4.9
F test	5.0	5.3	5.0	4.8	5.3	4.5	5.7	4.8	5.1

Table 4.19. *Empirical power in percent for the adaptive test and the F test for the treatment effect in a multicenter clinical trial. The total number of observations is indicated by n and the number of centers is indicated by c.*

	Distributions of the Errors								
Skewness	0	0	0	1	1	1	2	2	2
Kurtosis	1.8	3.0	6.0	4.2	5.4	8.4	11.4	12.6	15.6
	$n = 80,\ c = 5$								
Adaptive	44.2	36.3	41.4	47.3	44.5	46.0	48.8	56.9	52.9
F test	35.8	37.1	37.7	37.5	38.1	39.2	38.3	40.0	39.2
	$n = 80,\ c = 20$								
Adaptive	37.5	36.1	39.9	41.3	40.9	43.2	50.1	48.9	48.6
F test	35.4	36.9	37.7	37.8	37.5	38.8	38.6	39.0	39.9
	$n = 400,\ c = 20$								
Adaptive	53.0	36.5	43.0	51.6	46.2	47.3	64.9	60.5	56.0
F test	37.2	37.0	37.8	37.4	38.1	38.4	38.8	37.9	38.3

4.7 Summary and Recommendations

The adaptive test procedure described in Chapter 3 and the SAS macro used in this chapter were both designed to be flexible. The results of the simulation studies show that the adaptive test really does maintain its level of significance close to the nominal level, even for tests of subsets of coefficients in multiple regression models with highly correlated covariates and for tests for interaction effects in two-way layouts. In addition, for most simple models, the adaptive test has been shown to be more powerful than the traditional F test for nonnormal errors if the total number of observations equals or exceeds 20.

Because the test procedure can be used in place of the traditional F test for a subset of regression coefficients it can be used in many testing situations. The "awls" macro was designed so that the user only needs to specify the variable names for the variables that are used in the analysis. There are a few practical limitations on the use of the SAS macro. Depending on the size of the data set and the speed of the computer, the number of permutations may need to be restricted to a few thousand, but this should not be too much of a problem for most researchers. More importantly, the user should keep in mind that the data set used in the macro cannot contain missing values for any of the variables used in the macro. If missing values occur in any of the analysis variables, the user of the macro should make a SAS data set that has complete data for each observation. Missing values for variables that are not analyzed do not affect the performance of the macro. The macro relies heavily on PROC GLM to perform the adaptive test and the variables listed in the macro call are used in the MODEL statement or the CLASSES statement. Although interaction effects generated by SAS cannot be used, interaction effects created by the user can be specified in the "awls" macro.

In Chapters 2 and 4 we presented the results of many simulation studies that show that the adaptive test maintains its level of significance close to the nominal value. It would be helpful to have some theoretical results that would demonstrate this, but some large simulation studies with 100,000 data sets suggest that the size of the adaptive test may depart slightly from the nominal value. This is not too much of a practical problem because the empirical size of the adaptive test is often closer to the nominal value than that of the F test, which sometimes does not maintain its level of significance when the errors are nonnormal.

The power of the adaptive test has also been demonstrated by the simulation studies reported in previous chapters. These results consistently show that the adaptive test is considerably more powerful than the traditional test if the sample size is large, say, greater than 20, and the error distributions are skewed or have long tails. They also consistently show that there is very little loss in power if the adaptive test is used instead of the traditional test when the error distributions are approximately normal. When taken together, the simulation results show that the adaptive test can be recommended for most testing situations when the number of observations equals or exceeds 20. If there is a large number of parameters in the model, as in models used for the analysis of data from multicenter clinical trials, the number of observations may need to exceed 20. The adaptive test is not recommended for small data sets with fewer than 10 observations.

Chapter 5

An Adaptive Test for Paired Data

5.1 Some Examples of Paired Data

One of the more common statistical tests of significance is the test for paired data. Paired data can be obtained from experimental units when measurements are obtained before and after the administration of a treatment. Data from these pretest-posttest designs are sometimes called longitudinal data, or repeated measures data, or simply pretest-posttest data. With pretest-posttest designs we usually compute the difference as the posttest score minus the pretest score for each subject.

Paired data can also be obtained when the measurements are naturally paired. For example, data on the strength of the right hand and the strength of the left hand would be naturally paired if we had obtained both measurements on each subject. For the analysis of the paired data on hand strength we might compute the right-hand strength minus the left-hand strength for each person.

An example of paired data was given by Mitchell, Burroughs, and Beadles (1936), who investigated the protein of raw and roasted peanuts. In their experiment they randomly assigned one member of a pair of rats to a diet based on raw peanut protein, and the other member to a diet based on roasted peanut protein. After 5 weeks the differences in biological values for the $n = 10$ rat pairs were $\{6, 6, 9, 4, 5, 2, 2, 2, -18, 3\}$. Note the large number of positive differences and the large negative value for the ninth difference. These data were also analyzed by Snedecor and Cochran (1967, p. 96), who noted the outlier in the data set. In the analysis of these data we would usually perform a one-sided test of the null hypothesis that the mean biological values in the diet groups are equal, versus the alternative that the raw peanuts have a greater mean biological value.

Another example of paired data was given by Farber (1981), who listed data from identical twins reared in different families. One interesting comparison involved twins where one member of the pair was raised by a parent and the other by someone other than a parent. In Table E6 of Farber (1981) the IQ data were recorded for these pairs. We have listed the differences in the IQ scores between the twins, with each difference equaling the IQ for the member raised by someone other than a parent minus the IQ for the member raised by a parent. The differences for the $n = 33$ pairs of twins are $\{-4, 6, -13, 3, 14, 15,$

6, 3, 18, 14, 15, −1, −1, 1, 12, −12, 4, −9, −4, −11, 5, 11, −1, 5, 11, 4, 9, 19, 2, −5, −8, −3, 1}. We want to use the most powerful paired-sample test to test the null hypothesis that the children reared by others have the same average IQ scores as those reared by a parent. Since we will limit our discussion to one-sided alternatives in this chapter we will consider only the one-sided alternative that the children raised by others have higher IQ scores than those raised by a parent.

5.2 The Traditional Tests for Paired Data

The most commonly used test of significance for paired data is the paired t test, which is based on the differences between the members of the pairs. Let d_i denote the differences between the observations for the ith pair. The t test statistic for the test that the population mean difference equals zero is

$$t = \frac{\bar{d}}{s_d/\sqrt{n}},$$

where $\bar{d} = \sum_{i=1}^{n} d_i/n$ and $s_d^2 = \sum_{i=1}^{n} (d_i - \bar{d})^2/(n-1)$. The p-value for the t test is obtained from the t distribution with $n-1$ degrees of freedom. The t test assumes that the differences are normally distributed and is the most powerful one-sided test if the differences are normally distributed.

If the distribution of differences is not known to be normally distributed, a Wilcoxon signed-rank (SR) test is sometimes used. The SR test makes the less restrictive assumption that the distribution of differences is symmetric. We compute the SR test using the following procedure:

1. Rank the absolute values of the differences.
2. Attach the signs of the differences to the ranks of the absolute values of the differences.
3. Compute the sum S of the positive signed ranks.
4. Obtain the p-value from a table of the SR test. For larger samples with $n > 10$ compute the test statistic

$$\frac{S - n(n+1)/4}{\left\{n(n+1)(2n+1)/24\right\}^{1/2}}$$

and obtain the approximate p-value from a standard normal distribution. If some of the differences equal zero, or are tied, we need to modify the large sample approximation in the manner described by Hollander and Wolfe (1973, Chap. 3).

Neither the t test nor the SR test is adaptive in nature. The t test is the most powerful test when the differences are normally distributed. The SR test is the locally most powerful rank-based test if the differences follow the logistic distribution. However, not all distributions are normal or logistic. We would like an adaptive test that would be more powerful than either of these tests when the distribution of differences has long tails or short tails. We also want the adaptive test to be nearly as powerful as the t test with normally distributed data, and we want it to be nearly as powerful as the SR test if the differences are logistic.

5.3 An Adaptive Test for Paired Data

Adaptive tests for paired data are not new. Randles and Hogg (1973) suggested an adaptive test for the one-sample problem that used a selection statistic to choose one of three linear rank test statistics. The selection statistic was a measure of the tailweight for the absolute values of the differences. The rank test components of the adaptive test were the sign test, the SR test, and a rank test that used scores for light tails. Because the component tests were rank tests and because the selection statistics were independent of the ranks, the test maintained the nominal level of significance. It also had reasonable power when compared to the t test and the SR test.

Policello and Hettmansperger (1976) also proposed a one-sample adaptive test, but it did not maintain the nominal level of significance. Lemmer (1993) proposed two adaptive tests that can be used when the measurements are not symmetrically distributed. In Lemmer's adaptive test procedure the SR test is used if the empirical test is fairly symmetric, otherwise the sign test is used. This test appears to be useful for one-sample data that are not derived from paired data.

We will now describe an adaptive test for paired data that maintains the nominal level of significance and is more powerful than the t test and the SR test for nonnormal errors. This adaptive test uses the same overall approach as was used in the development of the test described in Chapter 3. However, this adaptive test for paired data is designed to make use of the symmetry of the distribution of the differences under the null hypothesis.

5.3.1 An Argument for Symmetry

Tests of significance used with paired data are usually quite different from those used with unpaired data. In the analysis of data from pretest-posttest designs, the null hypothesis is that the distribution of the pretest scores equals the distribution of the posttest scores. As a consequence, the distribution of the differences between the scores will be symmetric under the null hypothesis, even if the distributions of the pretest scores and the posttest scores are highly skewed. To understand why the differences are symmetrically distributed, consider any observed difference $d = x - y$, where x is the observation from the first member of a pair and y is the observation from the second member. Under the null hypothesis we have the same chance of observing x from the first member of the pair and y from the second member as we have of observing y from the first member and x from the second member. Consequently, under the null hypothesis, we are just as likely to observe $-d = y - x$ as we are to observe $d = x - y$. Since this is true for all pairs of observations (x, y) it can be shown, under the null hypothesis, that the distribution of differences is symmetric. For a more detailed argument see Lehmann (1975, Chap. 4).

Because the distribution of differences is assumed to be symmetric, the adaptive methods proposed in earlier chapters are not directly applicable. We will make modifications to our adaptive approach in order to take advantage of the symmetry of the population of differences.

5.3.2 The Doubling of Data

Let d_i denote the difference for the ith pair and let μ denote the mean of the differences. In this section we will describe an adaptive one-sided test of $H_0 : \mu = 0$ versus $H_a : \mu > 0$.

We have already noted that if we obtain a value of d_i for the ith observation, we will be just as likely to observe a difference of $-d_i$. Hence, we can construct an empirical distribution that is symmetric by creating an artificial difference $-d_i$ for each observation in the sample. That is, to construct a symmetric empirical distribution, we will "double" the sample differences by defining the double differences

$$d_{D,i} = \begin{cases} d_i & \text{for } i = 1, \ldots, n, \\ -d_{i-n} & \text{for } i = n+1, \ldots, 2n. \end{cases}$$

To determine the appropriate weights for the observations, we standardize them by using a robust estimate of the standard deviation. Let iqr_D be the interquartile range based on the doubled differences $\{d_{D,1}, \ldots, d_{D,2n}\}$. A robust estimate of the standard deviation, which was described by Fenstad, Kjaernes, and Walløe (1980), is $s_D = iqr_D / 1.349$. The standardized values for the doubled differences are then given by $z_{D,i} = d_i / s_D$ for $i = 1, \ldots, 2n$.

5.3.3 Smoothing and Weighting

We now use these $2n$ standardized values in a smoothing procedure to calculate a smooth c.d.f. Let $Z_D = \{z_{D,1}, \ldots, z_{D,2n}\}$. We will use a normal kernel smoother over the set of standardized doubled differences in order to obtain the smoothed c.d.f. at a point z as

$$\hat{F}_h(z; Z_D) = \frac{1}{2n} \sum_{i=1}^{2n} \Phi\left(\frac{z - z_{D,i}}{h}\right),$$

where $\Phi(\cdot)$ is the c.d.f. of the standard normal distribution and h is a smoothing constant used in the normal kernel. We use the same smoothing procedure as we used in Chapters 2 and 3, except that we double our data to obtain $2n$ observations. Consequently, we use $h = 1.587(2n)^{-1/3} = 1.26n^{-1/3}$ as the smoothing constant. Because the values have been standardized we do not need to include a factor in h for the standard deviation. Experimentation with several values of h has shown that the power and significance level of the test are not too sensitive to the choice of a smoothing constant. However, experimentation has also shown that very small values of h, which give little smoothing, often produce erratic weights.

For the ith standardized observation $z_{D,i}$, the value $\Phi^{-1}[\hat{F}_h(z_{D,i}; Z_D)]$ is the corresponding normal score. The adaptive test will use the weights

$$w_i = \frac{\Phi^{-1}\left[\hat{F}_h\left(z_{D,i}; Z_D\right)\right]}{z_{D,i}}$$

for $i = 1, \ldots, n$. If the differences are normally distributed, then $\Phi^{-1}[\hat{F}_h(z_{D,i}; Z_D)]$ should approximate $z_{D,i}$ for all observations, which will produce weights near one. In this situation the adaptive test should behave much like the t test. However, if the ith difference is an outlier, then $|z_{D,i}|$ will greatly exceed $|\Phi^{-1}[\hat{F}_h(z_{D,i}; Z_D)]|$ so that w_i will be small. Consequently, this weighting scheme will tend to downweight outliers.

The appropriate model for paired data is $d_i = \beta + \varepsilon_i$, and the usual one-sided t test is based on the test statistic for testing $H_0 : \beta = 0$ versus $H_a : \beta > 0$. With the adaptive approach we weight each observation to obtain the WLS model $w_i d_i = w_i \beta + w_i \varepsilon_i$. Substituting $y_i^* = w_i d_i$, $x_i^* = w_i$, and $\varepsilon_i^* = w_i \varepsilon_i$, we can write the transformed model as $y_i^* = x_i^* \beta + \varepsilon_i^*$. If we use OLS methods on the transformed data, we obtain the WLS estimate of β as

$$b = \frac{\sum_{i=1}^n x_i^* y_i^*}{\sum_{i=1}^n (x_i^*)^2} = \frac{\sum_{i=1}^n w_i^2 d_i}{\sum_{i=1}^n w_i^2}.$$

The adaptive test statistic is the traditional test statistic for testing $H_0 : \beta = 0$ versus $H_a : \beta > 0$ in the transformed model $y_i^* = x_i^* \beta + \varepsilon_i^*$. The adaptive test statistic is

$$t = \frac{b}{s^* \Big/ \sqrt{\sum_{i=1}^n (x_i^*)^2}} = \frac{b\sqrt{\sum_{i=1}^n w_i^2}}{s^*},$$

where s^* is the root mean squared error from the transformed data. Of course, since an adaptive method was used in the calculation of the test statistic, it will not follow a t distribution so we cannot obtain a p-value directly. To obtain a p-value we will need to use a permutation method.

5.3.4 A Permutation Method for Paired Data

To understand our permutation test, consider the ith pair of observations, which caused us to observe a difference of d_i. If we permute the observations in the ith pair, we obtain the difference $-d_i$, which, under the null hypothesis, is as likely to be observed as d_i. Thus, if we perform a permutation test, we randomly permute the observations within each pair so we are, in effect, randomly changing the signs of the differences. Consequently, the permutation test can be performed with paired data by calculating the test statistic for the 2^n sign changes for the n differences. Since it is not usually practical to compute all 2^n sign changes we usually take a large number R of these sign changes. Let E equal the number of sign changes that produce test statistics that exceed the observed test statistic t. We compute the p-value as $p = (E + 1)\big/(R + 1)$ and then reject the null hypothesis if $p < \alpha$, where α is the desired level of significance of the test.

It is important to understand the relationship between the sign changes and the weights. Since iqr_D is based on the doubled observations, the set $Z_D = \{d_1/s_D, \ldots, d_{2n}/s_D\}$ will contain the same elements after the sign changes. Hence, $\hat{F}_h(z; Z_D)$ will not change with sign changes. Since $\hat{F}_h(z; Z_D)$ is the c.d.f. of a symmetric distribution, a sign change for the ith difference will change the sign of the numerator and denominator of w_i. Thus, the weights will not change with sign changes. By constructing the test in this manner we ensure that the observed test statistic is just one of $R + 1$ equally likely test statistics under the null hypothesis.

Furthermore, some simplification is possible in the calculation of the p-value because many factors in the test statistic t do not change with sign changes. Because

$b = \sum_{i=1}^{n} w_i^2 d_i / \sum_{i=1}^{n} w_i^2$ the t test statistic can be written as

$$t = \frac{b\sqrt{\sum_{i=1}^{n} w_i^2}}{s^*} = \frac{\sum_{i=1}^{n} w_i^2 d_i}{s^* \sqrt{\sum_{i=1}^{n} w_i^2}}.$$

We note that $\sum_{i=1}^{n} w_i^2$ will not change with sign changes. We also note that $s^* = \sqrt{SSE^* / (n-1)}$, where

$$SSE^* = \sum_{i=1}^{n} w_i^2 d_i^2 - \left(\sum_{i=1}^{n} w_i^2 d_i \right)^2 \Big/ \sum_{i=1}^{n} w_i^2 .$$

Thus, because $\sum_{i=1}^{n} w_i^2 d_i^2$ will not change with sign changes, any increase in $\sum_{i=1}^{n} w_i^2 d_i$ will increase the numerator of t and decrease the denominator of t. Hence, any increase in $\sum_{i=1}^{n} w_i^2 d_i$ will increase t, so that the permutation test can be based on $\sum_{i=1}^{n} w_i^2 d_i$ instead of the test statistic t.

5.4 Using the SAS Macro

The adaptive test for paired data was used to analyze the $n = 10$ differences in biological protein values given by Mitchell, Burroughs, and Beadles (1936). We used the SAS macro for paired data listed in Appendix A. The macro "oneawls" is also available on the World Wide Web at the author's home page (*www.math.niu.edu/~ogorman/*). For this example, the differences, which were given a variable name of "diff," were stored in a SAS data set named "peanuts." To invoke the SAS macro we used the following SAS statement:

```
%oneawls(dataset=peanuts, y=diff, seed=143972,
        nperm=2000);
```

This macro produced the output given in Table 5.1. In this invocation of the macro we specified a starting seed for the random number generator and 2000 sign changes. The only required parameters are the "dataset" and the "y" variable. If the seed is not specified, it will be determined by the computer clock. The default value for "nperm" is 2000.

The smoothed c.d.f. of the $2n = 20$ doubled standardized differences in Z_D is given in Figure 5.1 along with the inverse c.d.f. of the standard normal. Since the interquartile range based on the doubled differences was 9.5, the robust standard deviation was 7.0422. Note that the ninth pair, which produced the smallest standardized difference, was given a standardized value of $z_{D,9} = -18/7.0422 = -2.556$. The smoothed c.d.f. at point $z_{D,9}$ was computed to be $\hat{F}_h(z_{D,9}; z_D) = .025$ so that $\Phi^{-1}(.025) = -1.96$. Consequently, the weight for the ninth difference was $w_9 = -1.96/-2.556 = 0.767$. The third pair produced the second-largest standardized difference and was given a weight of $w_3 = \Phi^{-1}[\hat{F}_h(z_{D,3}; z_D)] \,/\, z_{D,3} = 1.414 \,/\, 1.278 = 1.106$. The weights for the other observations were close to 1.0, so the ninth difference was the only difference that was greatly downweighted.

Based on $R = 2000$ permutations the p-value for the adaptive test was $p = .0585$, which is much less than the p-value of $p = .1969$ produced by the t test. The results of the

Table 5.1. *SAS output of macro "oneawls" for the peanut data.*

Adaptive weighted least squares analysis

Number of permutations = 2000 with first seed = 143972
SAS data set: peanuts

Variable name: diff

One-tailed p-value for adaptive test = 0.0585
Two-tailed p-value for adaptive test = 0.1169

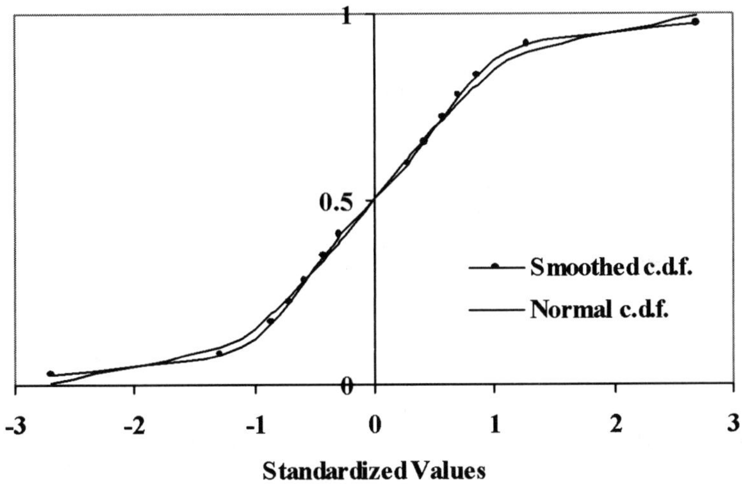

Figure 5.1. *The smoothed c.d.f. of the standardized doubled differences and the c.d.f. of the normal distribution for the peanut data.*

adaptive test suggest that raw peanuts were superior to roasted peanuts but, since we don't know the true difference, this does not demonstrate the superiority of the adaptive test.

This adaptive paired-sample test was also used to analyze the $n = 33$ differences in IQ scores from the twin data. For the doubled differences we obtained $iqr_D = 12$ and computed a robust standard deviation of $s_D = 8.9$. The p-value for the adaptive test was $p = .0150$, which is less than the $p = .0220$ produced by the t test and the $p = .0231$ produced by the SR test. As we saw in the last example, the adaptive test is somewhat more conclusive but, since we do not know the true difference, we cannot be certain that the adaptive test was really helpful. We will now describe a simulation study that was performed to evaluate the effectiveness of the adaptive procedure.

5.5 Size and Power Studies

5.5.1 Design of Simulation Studies

The simulation was designed to evaluate the size and power of the t test, the SR test, and the adaptive test. The simulation used six symmetric distributions with several sample sizes. To simulate data from short-tailed distributions, observations were generated from a uniform distribution with $\mu = 0$ and $\sigma = 1$ and from a beta distribution having parameters α and β with $\alpha = \beta = 2$, which was centered and rescaled so that $\mu = 0$ and $\sigma = 1$. These short-tailed distributions are shown in Figures 5.2 and 5.3. To simulate data from moderate tail length distributions, observations were generated from a standard normal and from a logistic distribution. The logistic distribution was also centered and rescaled so that $\mu = 0$ and $\sigma = 1$. These distributions are shown in Figures 5.4 and 5.5. To simulate data from one long-tailed distribution, observations were generated from a distribution that was a mixture of 90% from a standard normal and 10% from a normal outlier distribution with $\mu = 0$ and $\sigma = 10$. For another long-tailed distribution, observations were generated from a mixture of 70% from a standard normal and 30% from a normal outlier distribution with $\mu = 0$ and $\sigma = 10$. These long-tailed distributions are shown in Figures 5.6 and 5.7. Observations generated for the size simulation were obtained from the distributions directly, whereas those for a power simulation with a mean of δ were obtained by adding δ to a value obtained from one of the six distributions.

Sample sizes of $n = 10$, 20, and 50 were used in the Monte Carlo simulations. To evaluate the significance level of these tests, we generated 10,000 data sets from these six distributions and computed the number of rejections. All tests were one sided and used a significance level of $\alpha = .05$. Critical values for the t test were obtained from the t distribution with $(n - 1)$ degrees of freedom. The SR test used a critical value of 1.645 obtained from the normal distribution. For the adaptive test we used $R = 2000$ sign changes for simulations that used $n = 20$ and $n = 50$, but only $R = 500$ sign changes for simulations that used $n = 10$. To evaluate the powers of the four tests against a shift alternative, 10,000 data sets were generated from the six symmetric distributions and a constant δ was added to each observation. The constant δ was determined to give empirical powers near 50% for the t test.

5.5.2 Results of Simulation Studies

The empirical sizes of the adaptive test, the t test, and the SR test are given in Table 5.2 for the six symmetric distributions used to generate the differences in this simulation. Note that we do not generate the measurements for the two members of the pair to compute the difference. Instead, we generate the differences from a symmetric distribution. We used $\alpha = .05$ in a test of $H_0 : \beta = 0$ versus $H_a : \beta > 0$ with the model $d_i = \beta + \varepsilon_i$. The results clearly show that the adaptive test and the SR test maintain their level of significance. It should be observed that for $n = 10$ the t test does not maintain the level of significance of $\alpha = .05$ for long-tailed distributions. Even when $n = 20$, the t test does not maintain its significance level when the errors are generated from a 10% outlier distribution. Hence, the t test, which is a commonly used test for paired data, is not robust for validity when $n \leq 20$.

The empirical power is shown in Table 5.3 for the three tests with the same six distributions. The relative power of these tests is reported for one value of δ that was chosen

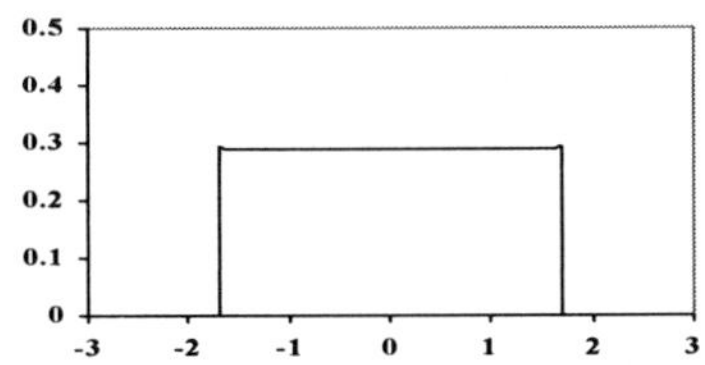

Figure 5.2. *The uniform distribution.*

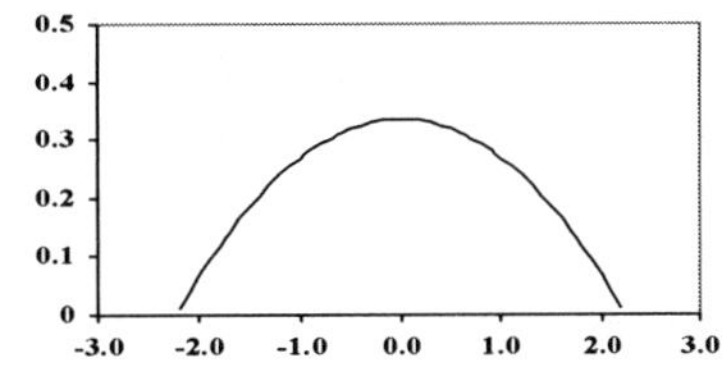

Figure 5.3. *The beta distribution.*

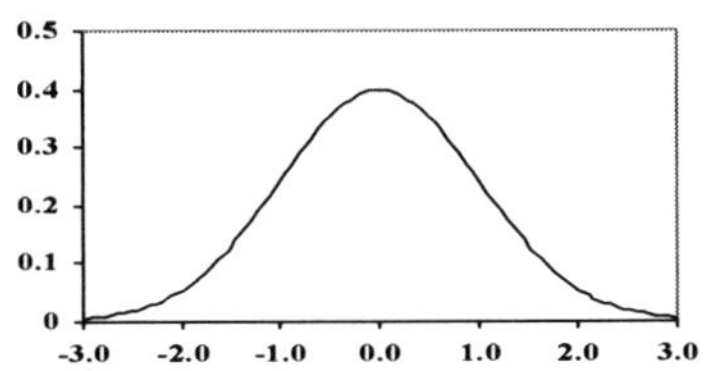

Figure 5.4. *The normal distribution.*

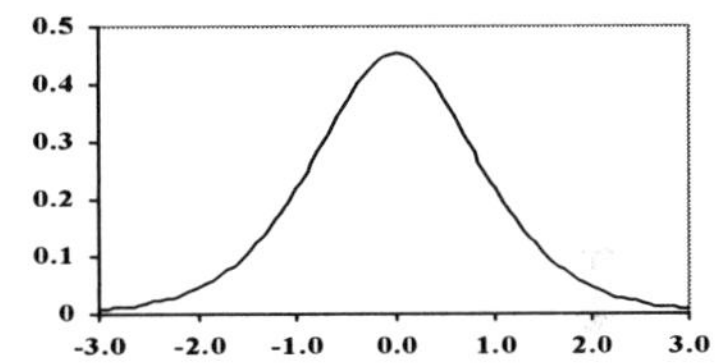

Figure 5.5. *The logistic distribution.*

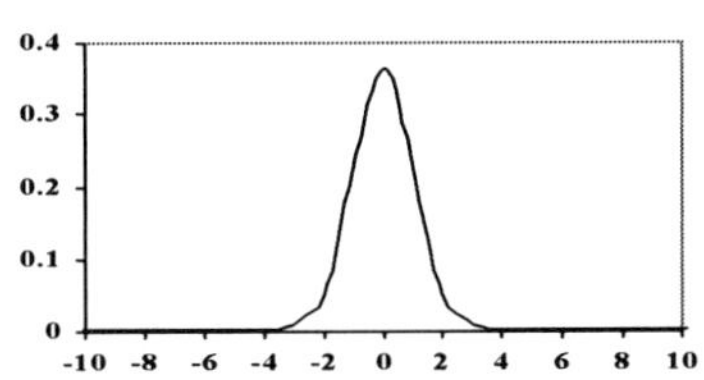

Figure 5.6. *The* 10% *outlier distribution.*

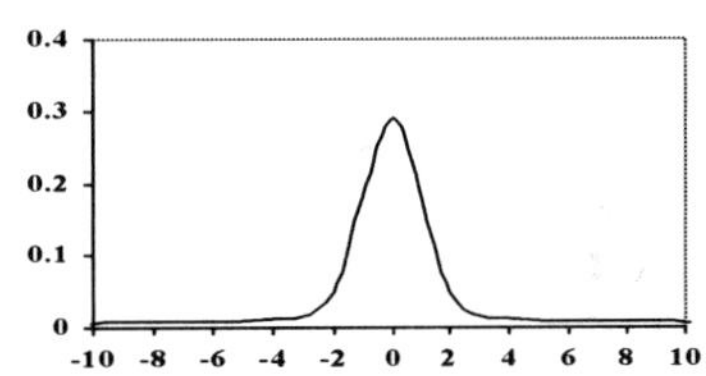

Figure 5.7. *The* 30% *outlier distribution.*

to give empirical powers near 50% for these one-sided tests. For $n = 10$ we note little difference between the power of the adaptive test and that of the SR test. For $n = 20$ and $n = 50$ the adaptive test is more powerful than the SR test for short- and long-tailed distributions. The t test is clearly inferior to the adaptive test for long-tailed distributions and has less than a 1% power advantage over the adaptive test when the differences are normally distributed.

We also investigated the performance of these tests when the shift model did not hold. Table 5.4 gives the empirical power for a small simulation study with $n = 20$ observations when an exponential random variate was added to the random variate generated from one of the six distributions. The exponential distribution had a mean value δ that was chosen to give empirical powers near 50%. The relative performance of these tests is similar to that observed in Table 5.3 for the shift model. We observe that the t test has low power for long-tailed distributions and that the adaptive test is more powerful than the SR test for long- and short-tailed distributions.

Table 5.2. *The empirical size in percent for a test of $H_0 : \mu = 0$ versus $H_a : \mu > 0$ using a significance level of $\alpha = .05$.*

	Distribution					
	Uniform	Beta	Normal	Logistic	10% Outliers	30% Outliers
			$n = 10$			
Adaptive test	5.0	5.1	4.6	5.1	4.7	5.1
t test	5.2	5.1	4.7	4.9	2.8	3.2
SR test	5.3	5.3	4.9	5.4	5.0	5.4
			$n = 20$			
Adaptive test	5.5	4.9	4.9	4.7	5.0	5.1
t test	5.4	5.0	4.9	4.8	3.4	4.7
SR test	5.3	4.7	4.8	4.6	4.7	4.9
			$n = 50$			
Adaptive test	4.8	5.4	4.4	4.8	4.7	4.8
t test	4.6	5.5	4.4	4.8	4.2	4.8
SR test	4.8	5.7	4.5	4.8	4.7	4.8

Table 5.3. *The empirical power in percent for a test of $H_0 : \mu = 0$ versus $H_a : \mu > 0$ using a significance level of $\alpha = .05$.*

	Distribution					
	Uniform	Beta	Normal	Logistic	10% Outliers	30% Outliers
			$n = 10$			
Adaptive test	41.2	42.7	45.2	47.2	63.0	52.3
t test	40.8	42.9	45.9	47.6	47.6	32.9
SR test	39.6	41.7	45.9	49.2	62.6	57.6
			$n = 20$			
Adaptive test	49.4	47.5	46.6	50.1	70.2	70.9
t test	46.0	46.3	47.1	49.7	39.6	27.2
SR test	42.6	41.9	45.1	50.6	65.2	64.0
			$n = 50$			
Adaptive test	58.0	52.2	48.7	50.3	75.7	86.1
t test	48.6	48.9	49.2	49.3	29.6	23.5
SR test	45.4	44.7	48.0	51.8	70.1	72.8

5.6 Recommendations for Paired Data Tests

Based on the results of this Monte Carlo study, there appears to be no reason to use the t test for studies with $n \geq 10$. The t test has relatively low power for long-tailed distributions and only a small power advantage over the adaptive test for the normal distribution. The poor finite sample performance of the t test relative to the SR test has also been noted by Blair and

Table 5.4. *The empirical power in percent for a test of $H_0 : \mu = 0$ versus $H_a : \mu > 0$ using a significance level of $\alpha = .05$ when an exponential random variate with a mean of δ was added to the value generated from a symmetric distribution with a mean of zero.*

	Distribution					
	Uniform	Beta	Normal	Logistic	10% Outliers	30% Outliers
			$n = 20$			
Adaptive test	46.6	43.9	43.2	44.1	58.7	58.0
t test	42.6	43.4	44.9	44.8	36.3	24.8
SR test	38.7	39.5	42.3	44.4	54.6	50.1

Higgins (1985). Consequently, only the SR test and the adaptive test can be recommended with paired data for $n \geq 10$. The results show that either the SR test or the adaptive test is a reasonable choice for studies with about 10 observations. For studies with $n < 10$ the t test may be more powerful than the other tests, but the relative performance of these tests was not investigated in these simulation studies.

The power results in Table 5.3, for $n = 20$ and $n = 50$ with a constant treatment effect and the results in Table 5.4 with an exponential treatment effect show that the power of the adaptive test approximates the power of the t test and the SR test for normal and logistic distributions and that the power of the adaptive test exceeds the power of the other tests for long- and short-tailed distributions. Consequently, the adaptive test is recommended for studies with $n \geq 20$.

In addition, the adaptive test is not difficult to perform. The analysis of a single data set takes very little time with this macro, even with 2000 sign combinations. The number of sign combinations can be set much higher if there is some desire to obtain a more precise estimate of the p-value.

Chapter 6
Adaptive Confidence Intervals

6.1 The Relationship between Confidence Intervals and Tests

Confidence intervals are used in many areas of the sciences to give researchers some idea of the precision of their estimates. These intervals are often used in circumstances where a test of significance is simply not appropriate. For example, a researcher may randomly assign animals to a treatment group and a control group in order to estimate the magnitude of a treatment effect. If the treatment is known to have some effect, then a confidence interval, rather than a test of significance, is more appropriate because the confidence interval gives some idea about the magnitude of the effect. In this chapter we will develop procedures for computing adaptive confidence intervals. We would like these adaptive confidence intervals to be narrower than the traditional ones when the error distribution is nonnormal and nearly as narrow when the error distribution is normal.

In contrast to the large amount of literature on adaptive testing, there have been few articles on adaptive confidence intervals. An adaptive approach to confidence interval construction was taken by Clarke (1994), who evaluated the performance of several adaptive confidence intervals for the mean of a symmetric distribution. Using a WLS approach, O'Gorman (2001c) proposed a method of constructing adaptive confidence intervals for a single parameter in a linear model. Although the method appeared promising, it is not as effective as the method that will be described in this chapter.

Our objective in this chapter is to construct adaptive confidence intervals that maintain their coverage probability and are usually narrower than the traditional confidence intervals. If we use a method to produce an adaptive 95% confidence interval, we should find, after taking many samples from the same population, that 95% of the confidence intervals cover the population parameter. Consequently, after we have developed our adaptive confidence intervals, we will first check to see if they maintain their coverage probability and then compare the average width of the adaptive intervals to the average width of the traditional confidence intervals.

Fortunately, because tests of significance are closely related to confidence intervals, we can use many of the adaptive testing procedures in the development of adaptive confidence intervals. To illustrate the relationship between confidence intervals and tests of significance, suppose that we have taken one sample from a population in order to compute a $100(1-\alpha)\%$ confidence interval for the population mean μ. We recall that if a point μ_0 is in the confidence interval, then we will fail to reject the null hypothesis $H_0 : \mu = \mu_0$ when tested against the two-sided alternative $H_a : \mu \neq \mu_0$ using a level of significance of α. Consequently, it is possible to construct an adaptive confidence interval by finding all the values of μ_0 in an adaptive test of $H_0 : \mu = \mu_0$ that would lead to a failure to reject the null hypothesis. The relationship between confidence intervals and tests is explained in Bickel and Doksum (1977, Chap. 5). In this chapter we will confine our discussion to two-sided confidence intervals, but the same basic approach could be used with one-sided intervals.

In practice, a confidence interval may be difficult to determine because of the large number of tests that would be required. An additional difficulty is that the adaptive tests described in the previous chapter require a permutation method to estimate a p-value. Because these permutation methods are computationally intensive we will use an iterative search method for the lower and upper limits in order to keep the amount of calculation at a reasonable level. This search algorithm will be described in the next section.

6.2 Using Permutation Tests to Compute Confidence Intervals

In this section we will describe a permutation method that can be used to compute confidence intervals. Although we could try many values of μ_0 in a two-sided test of significance, an easier way to find the confidence interval is to use a procedure described by Garthwaite (1996), which is based on the search process originally described by Robbins and Monro (1951).

In the following two sections we will construct adaptive versions of some of the most popular confidence intervals, including an adaptive confidence interval for the slope in a simple linear regression and another for the difference between two means. In this section we will describe a general search method for the confidence interval for one parameter in a linear model. To develop the confidence interval we will specify the linear models corresponding to the null and alternative hypotheses in the same manner as we specified in Chapter 3. The complete model is given by

$$Y = X_C \beta_C + \varepsilon,$$

where Y is an $n \times 1$ column vector of observations, X_C is an $n \times (p+1)$ model matrix, β_C is a $(p+1) \times 1$ vector of parameters, and ε is an $n \times 1$ vector of error terms. We next partition X_C and β_C to correspond to the null and alternative hypotheses. Let β_A be the last parameter in β_C and let β_R be the $p \times 1$ vector corresponding to the first p parameters in β_C. We partition X_C similarly, with X_R containing the first p columns of X_C and X_A containing the last column of X_C. With this notation the model can be written as

$$Y = X_R \beta_R + X_A \beta_A + \varepsilon.$$

Now suppose that we want a $100(1-\alpha)\%$ confidence interval for β_A. A value c will be in the confidence interval if the p-value for the upper-tailed test of $H_0 : \beta_A = c$ versus

$H_a : \beta_A > c$ is greater than $\alpha/2$ and the p-value for the lower-tailed test of $H_0 : \beta_A = c$ versus $H_a : \beta_A < c$ is greater than $\alpha/2$. Although we could use many values of c in these tests to determine the $100(1-\alpha)\%$ confidence interval for β_A, we will use a search method to find the upper and lower limits of the interval.

The iterative method proposed by Garthwaite (1996) requires a separate search for the upper and lower limits based on initial guesses of the limits of the interval. Let L_i be the estimate of the lower confidence limit (β_L) at the ith iteration and let U_i be the estimate of the upper confidence limit (β_U) at the ith iteration. In the search for the lower confidence limit the estimate is adjusted after each iteration by comparing the test statistic computed from the unpermuted data to the test statistic obtained by permuting the data once. After each iteration the lower limit is adjusted up or down by an amount that decreases with the number of iterations. The upper limit is estimated in a similar fashion. Generally, only a few thousand iterations are needed to obtain reasonably accurate estimates of each limit.

We begin the iterative procedure by estimating the lower limit of a $100(1-\alpha)\%$ confidence interval. Suppose that at the ith iteration we have a lower limit estimate of L_i. We then use the adaptive test procedure to compute the unpermuted t test statistic for $H_0 : \beta_A = L_i$, which is denoted by t^*. Next, one permutation of X_A is performed and the permutation test statistic (t^{perm}) is computed. These test statistics are used to update the estimate of β_L using the following formulas:

$$L_{i+1} = \begin{cases} L_i + k_i\left(\alpha/2\right)/i & \text{if } t^{perm} < t^*, \\ L_i - k_i\left(1-\alpha/2\right)/i & \text{if } t^{perm} \geq t^*, \end{cases}$$

where k_i is a steplength value. For $\alpha = .05$, Garthwaite (1996) recommended using $k_i = 17.46(\hat{B}_A - L_i)$, where $\hat{B}_A$ is a robust estimator of B_A. If we were computing the confidence interval for the mean of a symmetric population, we would set $\hat{B}_A$ equal to the median. If we were computing the confidence interval for one parameter in a linear model, we would use an adaptive estimator of β_A that will be described in Chapter 7. This adaptive estimator is obtained by using the complete model to determine the weights to be used with the WLS regression. The choice of an estimator of β_A is not very important because it only influences the step size used in the iterative process.

We begin the iterative procedure to calculate a 95% confidence interval by using the limits of the traditional 95% confidence interval as the initial estimates of β_L and β_U. Following the recommendation of Garthwaite (1996), we did not start the iterations with $i = 1$ but instead with $i = \lfloor .3(2-\alpha/2)/(\alpha/2)\rfloor$, where $\lfloor x \rfloor$ is the greatest integer less than or equal to x. For 95% confidence intervals we use $\alpha = .05$ so we will start the iterations with $i = 23$ to avoid making large corrections to the estimate. The iterative method was used to adjust the values of L_i until the convergence criteria were satisfied. We performed the steps in the iteration until $i = 5999$ and then considered successive blocks consisting of 1000 iterations. Two criteria were used to determine when the iterative process could stop. The iterations stopped after a block of 1000 iterations if the range of the L_i within the block was less than $.01(\hat{B}_A - L_l)$, where l was the last value of i in the block, and if the number of times that t^{perm} equaled or exceeded t^* was between $\left[1000\left(\alpha/2\right)-5\right]$ and $\left[1000\left(\alpha/2\right)+5\right]$. The first criterion ensured that the iterative estimates were quite stable. The second criterion makes sense because, for large values of i, t^{perm} should exceed t^* for approximately $1000\left(\alpha/2\right)$ iterations. This second criterion may not be absolutely essential, but it is used to help ensure that the correct limit has been found. Successive blocks of 1000 iterations were used until both convergence criteria were satisfied. This stopping rule required at least 6999 iterations but seldom required more than 12,999 iterations.

This same general approach was used to compute the upper limit. The updating formula is

$$U_{i+1} = \begin{cases} U_i - k_i \left(\alpha/2\right)/i & \text{if } t^{perm} > t^*, \\ U_i + k_i \left(1 - \alpha/2\right)/i & \text{if } t^{perm} \le t^*, \end{cases}$$

with a steplength value of $k_i = 17.46(U_i - \hat{B}_A)$. The iterations were started at $i = 23$ and the stopping rule for the upper limit was similar to the stopping rule for the lower limit.

Some understanding of the method can be obtained by considering how the estimates behave for large values of i when the estimates vary little from iteration to iteration. Consider the behavior of the estimate of the upper limit from iteration i to iteration $i + N$. If i is large, then k_i/i will be nearly constant over the next N steps. Hence, if U_i is near β_U, then, in the next N steps, the condition $t^{perm} \le t^*$ will occur approximately $(\alpha/2)\,N$ times, which will increase the estimate by approximately $k_i \left[\left(1 - \alpha/2\right)/i\right](\alpha/2)\,N$. Also, in those N steps the condition $t^{perm} > t^*$ will occur approximately $(1 - \alpha/2)\,N$ times, which will decrease the estimate by approximately $k_i \left[\left(\alpha/2\right)/i\right](1 - \alpha/2)\,N$. Consequently, if U_i is near β_U after a large number of steps, the amount of decrease will approximate the amount of increase so that U_{i+N} will remain near β_U. For a more detailed explanation of the search method and for several examples of the procedure the reader should refer to Garthwaite (1996) and to Garthwaite and Buckland (1992).

By using this iterative method we can find a nonadaptive confidence interval for any single parameter in a linear model. We now turn our attention to using this approach to construct adaptive confidence intervals. In section 6.3 we will apply this iterative procedure to find the adaptive confidence interval for the mean of a symmetric distribution. In section 6.4 we will construct an adaptive confidence interval for one parameter in a linear model.

6.3 An Adaptive Confidence Interval for the Mean of a Symmetric Distribution

6.3.1 The Adaptive Confidence Interval

In this section we will describe how to compute an adaptive confidence interval for the mean of a symmetric distribution. The iterative method described in the previous section will be used, with some modifications to take advantage of the symmetry of the distribution. We will assume that the appropriate model is $y_i = \beta + \varepsilon_i$, where y_i is the observed random variable and ε_i is the random error for the ith observation. Although the error distribution will be assumed to be symmetric about a mean value, we will not assume that it is normal.

In order to find the confidence interval, we need to determine the range of values that, when used in a test of significance, would cause us to accept the null hypothesis. Because we are assuming that the distribution is symmetric we must use a test that assumes that the population is symmetric about some mean value. In the adaptive test for paired data that we described in section 5.3 we assumed, under the null hypothesis, that the population was symmetric about zero. We can generalize this test for paired data so that it can be used to test the null hypothesis that the population is symmetric about a nonzero mean. This test is generalized by subtracting the hypothesized mean value from each observation to obtain the adjusted data, which will, under the null hypothesis, be symmetrically distributed about zero. Thus, by suitable adjustment of the data, we can use the adaptive test for paired data to find the confidence interval for the mean.

We begin our iterative method by using $L_i = \bar{y} - t_{\alpha/2,n-1}s/\sqrt{n}$ as the initial estimate of the lower limit, where $\bar{y}$ and s are the sample mean and standard deviation of $\{y_1, \ldots, y_n\}$. We start the iterative process with $i = 23$, as suggested by Garthwaite (1996), in order to avoid the large corrections to the estimate in the first few iterations. At each iteration we test $H_0 : \beta = L_i$ against $H_a : \beta < L_i$. In order to use the adaptive test for paired data, which always uses zero as the hypothesized mean under the null hypothesis, we will need to adjust the data and the model. We let $\beta_{ADJ} = \beta - L_i$ so that $\beta = \beta_{ADJ} + L_i$. Hence, the model can be written, in matrix form, as $Y = 1\beta + \varepsilon = 1(\beta_{ADJ} + L_i) + \varepsilon$, where 1 is a column of n ones. We then adjust the y-values to obtain $Y_{ADJ} = Y - 1L_i$ so that we can write the adjusted model as $Y_{ADJ} = Y - 1L_i = 1(\beta_{ADJ} + L_i) + \varepsilon - 1L_i = 1\beta_{ADJ} + \varepsilon$. Testing the null hypothesis $H_0 : \beta = L_i$ against $H_a : \beta < L_i$ in the original model is equivalent to testing $H_0 : \beta_{ADJ} = 0$ against $H_a : \beta_{ADJ} < 0$ with the adjusted model.

Consequently, to compute an adaptive confidence interval for β we first compute $y_{ADJ,i} = y_i - L_i$ for $i = 1, \ldots, n$. We then use the procedures for the adaptive test for paired data described in section 5.3 to compute the adaptive test statistic. This involves doubling the adjusted data, computing the empirical c.d.f. of the doubled data, computing the weights, and then computing the test statistic t^*.

Note that, under $H_0 : \beta = L_i$, the distribution of errors in the adjusted model is symmetric, which implies that $f(y_{ADJ,i}) = f(-y_{ADJ,i})$ for $i = 1, \ldots, n$. Thus, any random sign combination of the values in $\{y_{ADJ,1}, \ldots, y_{ADJ,n}\}$ is as likely to be observed as any other sign combination. For this iterative method we generate, at random, sign combinations for the n values in $\{y_{ADJ,1}, \ldots, y_{ADJ,n}\}$ and then compute the adaptive t test statistic t^{perm} based on those sign combinations. As we indicated in section 5.3, the weights will not change with sign changes. After comparing t^{perm} to t^* we adjusted the lower limit using the updating formula

$$L_{i+1} = \begin{cases} L_i + k_i\left(\alpha/2\right)/i & \text{if } t^{perm} < t^*, \\ L_i - k_i\left(1 - \alpha/2\right)/i & \text{if } t^{perm} \geq t^*. \end{cases}$$

This iterative process was continued until the convergence criteria specified in section 6.2 were satisfied.

We used a similar procedure for the upper limit using $U_i = \bar{y} + t_{\alpha/2,n-1}s/\sqrt{n}$ as the starting value. The data were adjusted using $y_{ADJ,i} = y_i - U_i$ for $i = 1, \ldots, n$, and the adaptive test was used to compute the test statistic t^*. At each step in the iterative procedure one sign combination was generated and the adaptive test statistic t^{perm} was computed. Then we obtained the next estimate of the upper limit by using the updating formula

$$U_{i+1} = \begin{cases} U_i - k_i\left(\alpha/2\right)/i & \text{if } t^{perm} > t^*, \\ U_i + k_i\left(1 - \alpha/2\right)/i & \text{if } t^{perm} \leq t^*. \end{cases}$$

We continued to update the upper limits until the convergence criteria were satisfied.

6.3.2 An Example

In a study of the effects of several diets on $n = 20$ men who were diagnosed with hypercholesterolemia, Anderson et al. (1984) recorded the change in high-density lipoprotein (HDL) cholesterol concentrations. The initial and final measurements (in milligrams per deciliter) were recorded and the differences were calculated as the final HDL concentration minus the initial HDL concentration. These differences are shown in the dot plot in Figure 6.1. While most of the differences were between -8.3 mg/dL and $+1.4$ mg/dL, there was one large positive outlier with a value of $+10.8$ mg/dL and a large negative outlier with a value of -13.3 mg/dL.

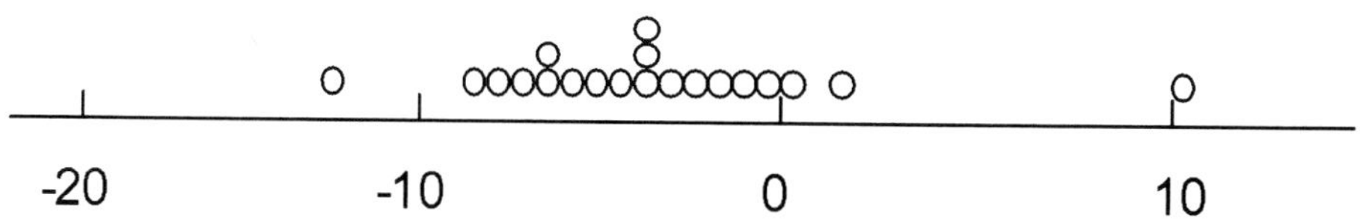

Figure 6.1. *Plot of differences in HDL cholesterol concentrations for* $n = 20$ *male subjects with hypercholesterolemia.*

Using the traditional method, we computed a 95% confidence interval of $\bar{y} \pm t_{\alpha/2,n-1}s/\sqrt{n} = [-5.1, \ -0.7]$. When the adaptive confidence interval method was used we obtained a 95% confidence interval of $[-4.7, \ -1.4]$. The adaptive interval was considerably narrower than the traditional interval because the two outliers were downweighted. To satisfy the convergence criteria the adaptive method required 8999 iterations for the lower limit and 9999 iterations for the upper limit. For the adaptive lower limit, the large positive difference was given a weight of 0.52 for the last iteration, the large negative difference was given a weight of 0.74, and the weights of the other observations varied over the narrow range of 0.84 to 0.93. For the upper limit the large positive difference was given a weight of 0.66 for the last iteration, the large negative difference was given a weight of 0.63, and the other weights were in the range 0.93 to 1.12. Thus, we see that the two outliers were downweighted in the analysis, which produced an adaptive confidence interval that was narrower than the traditional interval.

A SAS macro listed in Appendix A can be used to compute the adaptive 95% confidence limits for the population mean. If we have the HDL concentrations in a SAS data set called "diets" and derive a variable "hdl" that is the final HDL concentration minus the initial concentration, then the SAS macro "cione" can be invoked with the following statement:

```
%cione(dataset=diets, y=hdl, seed=124943, maxiter=8000);
```

This SAS macro uses the iterative procedure to find the limits obtained after "maxiter" iterations. The convergence criteria that have been described for the simulation studies are not used in the SAS macro. Because the convergence criteria are so conservative they cause the iterative procedures to continue longer than is really necessary and, because the macro does not execute quickly, it seemed more practical to allow the user to specify the number of iterations that should be used. Since the macro prints the limits after each set of

1000 iterations, the user should be able to verify that the limits have converged after a few thousand iterations. The macro runs slowly because the values of y_{ADJ} change slightly with each iteration, which then requires the smoothing of the c.d.f. in the weighting procedure with each iteration. It is a good idea to run this macro with "maxiter" set to a small value, say, 10, to determine if it is running properly before using a large value for the maximum number of iterations.

6.3.3 Coverage Probabilities and Average Widths

A Monte Carlo simulation study was used to determine if the adaptive confidence interval for the mean of a symmetric population maintained its coverage probability and if the adaptive intervals were narrower, on average, than the traditional confidence intervals. The simulation study used the same six distributions that were used in Chapter 5: the uniform, beta, normal, logistic, 10% outlier, and 30% outlier distributions. We then added one to those variates so that they were symmetric about one. This was done because we wanted to use a nonzero mean to be assured that the confidence intervals would behave properly with distributions that are not symmetric about zero. For each of these six symmetric distributions we performed three simulation studies with sample sizes of $n = 10,\ 20$, and 50.

For each distribution and sample size, 1000 data sets were generated. For each data set the iterative search method continued until the convergence criteria were satisfied in order to compute the lower and upper limits of the adaptive confidence interval. The average width was recorded over the 1000 data sets along with a count of the number of intervals that included the population mean.

Table 6.1 contains the empirical coverage probabilities of the adaptive and the traditional 95% confidence intervals for the six distributions and three sample sizes. With 1000 data sets the standard error of these coverage probabilities is approximately 0.7%. The results show that the adaptive confidence intervals have empirical coverage probabilities that approximate 95%. In addition, the results suggest that the traditional confidence interval covers the population mean too often when long-tailed distributions are used with $n \leq 20$ observations. Thus, the coverage probabilities of the adaptive confidence interval appear to be closer to the nominal value than those of the traditional confidence interval.

We now turn our attention to the widths of these confidence intervals. The average widths of the adaptive and traditional confidence intervals are given in Table 6.2, where the average is computed over the 1000 data sets used in each simulation study. In addition to tabulating the average widths, we have tabulated the relative efficiencies of the adaptive confidence interval to the traditional confidence interval. The relative efficiency is defined as

$$\text{Relative efficiency} = \left(\frac{\text{Average width of OLS intervals}}{\text{Average width of adaptive WLS intervals}} \right)^2 .$$

Since the widths of the intervals are roughly proportional to $1/\sqrt{n}$, the relative efficiency is a measure of how, if we use the adaptive confidence interval, the sample size can be reduced to achieve the same width that would have been obtained by using the traditional confidence interval. A rough interpretation of a relative efficiency of 2.0 is that, by using an adaptive confidence interval, the sample size can be divided by 2.0 to obtain roughly the same average width as the traditional interval. The results in Table 6.2 show that there is

Table 6.1. *Empirical coverage probabilities in percent of the adaptive and traditional 95% confidence intervals for the mean of a symmetric distribution.*

	Distribution					
	Uniform	Beta	Normal	Logistic	10% Outliers	30% Outliers
			$n = 10$			
Adaptive	94.8	93.7	94.7	95.1	94.2	94.8
Traditional	94.5	93.1	95.1	95.1	97.0	97.8
			$n = 20$			
Adaptive	93.4	96.1	94.4	96.3	95.3	95.6
Traditional	93.1	95.9	94.4	96.4	98.2	96.5
			$n = 50$			
Adaptive	95.3	93.9	94.6	95.0	95.0	95.1
Traditional	95.0	94.3	95.0	94.9	96.2	95.0

Table 6.2. *The average widths of the adaptive and traditional confidence intervals for the mean of a symmetric distribution and the relative efficiencies of the adaptive confidence interval to the traditional confidence interval.*

	Distribution					
	Uniform	Beta	Normal	Logistic	10% Outliers	30% Outliers
			$n = 10$			
Adaptive	1.42	1.41	1.39	1.36	3.68	6.84
Traditional	1.40	1.40	1.39	1.37	3.88	7.13
Relative Eff.	0.97	0.99	1.00	1.01	1.11	1.09
			$n = 20$			
Adaptive	0.92	0.93	0.93	0.91	1.47	3.33
Traditional	0.92	0.93	0.92	0.91	2.75	4.95
Relative Eff.	1.01	1.00	0.99	1.01	3.49	2.22
			$n = 50$			
Adaptive	0.51	0.55	0.57	0.55	0.66	1.01
Traditional	0.57	0.57	0.56	0.56	1.77	3.09
Relative Eff.	1.22	1.06	0.99	1.04	7.29	9.40

little difference in average width between the adaptive and traditional confidence intervals for the beta, normal, and logistic distributions. However, the results also clearly show that, with long-tailed distributions, the adaptive confidence interval is much narrower, on average, than the traditional confidence interval. Consequently, for the analysis of paired data, the adaptive confidence interval appears to be superior to the traditional confidence interval.

We computed the average widths of the confidence intervals by averaging the widths over the 1000 data sets used in the simulations. Is it possible that, with the averaging of the widths of these confidence intervals, the average does not reflect the typical width? To show that the average width does fairly describe the widths of these confidence intervals, a scatterplot of the widths of the adaptive and the traditional confidence intervals is given in Figure 6.2 for the data sets with $n = 20$ observations generated from a 10% outlier distribution. The scatterplot clearly shows that the adaptive method often produced shorter intervals than the traditional method, whereas the traditional methods never produced intervals much shorter than the adaptive intervals.

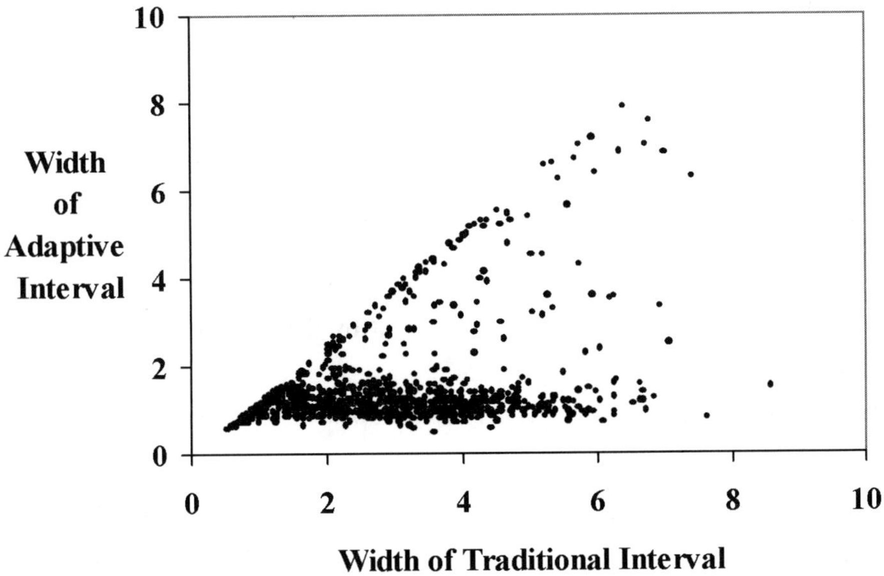

Figure 6.2. *A scatterplot of the widths of the adaptive and the traditional intervals for* 1000 *data sets of size* $n = 20$ *generated from a* 10% *outlier distribution.*

In contrast, Figure 6.3 gives the scatterplot of the widths of the adaptive and traditional confidence intervals for 1000 data sets with $n = 20$ observations generated from a normal distribution. This scatterplot shows that, for most data sets, there is close agreement between the widths for the two confidence intervals. Consequently, if the errors are normally distributed, there is little loss, or gain, in precision if the adaptive method is used to construct confidence intervals. The results shown in Figures 6.2 and 6.3 demonstrate that the average widths in Table 6.2 fairly represent the overall performance of the traditional and adaptive confidence intervals.

6.4 An Adaptive Confidence Interval for One Parameter in a Linear Model

We can also use the iterative procedure that we described in section 6.2 to compute an adaptive confidence interval for any single parameter in a linear model. To calculate the confidence interval we again rely on the duality between the confidence interval and the

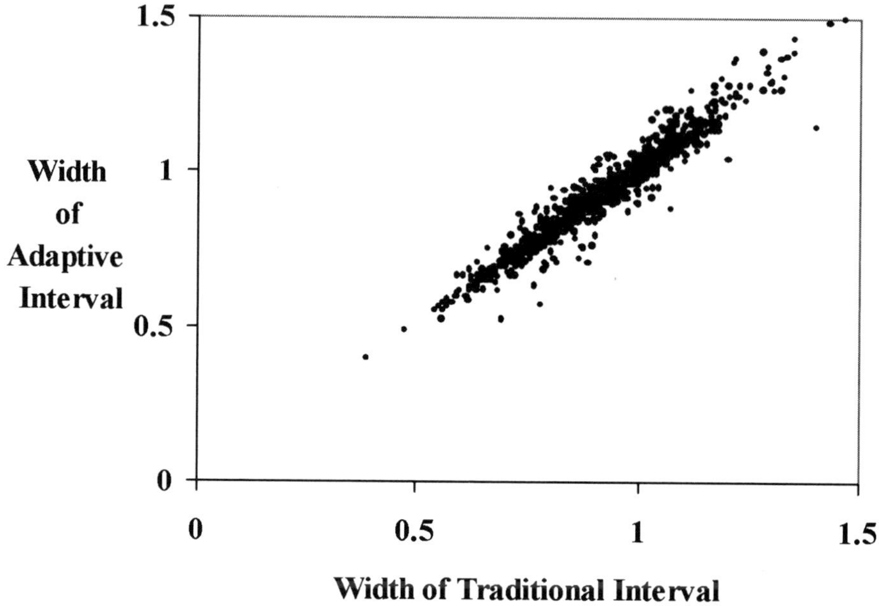

Figure 6.3. *A scatterplot of the widths of the adaptive and the traditional intervals for* 1000 *data sets of size* $n = 20$ *generated from a normal distribution.*

corresponding test of significance. In Chapter 3 we tested $H_0 : \beta_A = 0$ versus $H_a : \beta_A \neq 0$ in the model

$$Y = X_R\beta_R + X_A\beta_A + \varepsilon,$$

where X_R was an $n \times (q+1)$ matrix, β_R was a $(q+1) \times 1$ vector of parameters, X_A was an $n \times (p-q)$ matrix, β_A was a $(p-q) \times 1$ vector of parameters, and ε was an $n \times 1$ vector of errors. We included an intercept in each model by letting the first column of X_R be a column of ones. In this chapter we will restrict β_A to a single parameter so that $q = p - 1$. Consequently, β_A will be a 1×1 vector and β_R will be an $n \times [(p-1)+1] = n \times p$ vector. To use the relationship between confidence interval and tests of significance we note that, if a value c is in the $100(1-\alpha)\%$ confidence interval for β_A, then the test of $H_0 : \beta_A = c$ versus $H_a : \beta_A \neq c$ should fail to reject the null hypothesis using a significance level of α.

To perform a test of $H_0 : \beta_A = c$, it will be convenient to let $\beta_{ADJ} = \beta_A - c$. Using this notation, the full model can be written as

$$Y = X_R\beta_R + X_A(\beta_{ADJ} + c) + \varepsilon.$$

If we use the adjusted y-values $Y_{ADJ} = Y - cX_A$, the model becomes

$$Y_{ADJ} = Y - cX_A = X_R\beta_R + X_A\beta_{ADJ} + \varepsilon.$$

Since $\beta_{ADJ} = \beta_A - c$ we can see that, with the adjusted y-values, testing $H_0 : \beta_{ADJ} = 0$ is equivalent to testing $H_0 : \beta_A = c$ with the unadjusted data. Consequently, to obtain the least squares estimates of the elements in β_R under the null hypothesis $H_0 : \beta_A = c$, we will compute the adjusted y-values and then use the reduced model

$$Y_{ADJ} = X_R\beta_R + \varepsilon.$$

The residuals from this model are used to determine the weights that should be given to each of the observations.

The weighting procedure that will be used to compute adaptive confidence intervals is identical to that described in Chapter 3. First, the studentized deleted residuals d_i are computed for each observation. Then the smoothed c.d.f. of these residuals is computed using a normal kernel with a bandwidth of $h = 1.587n^{-1/3}$. The smoothed c.d.f. is used to estimate the median and the studentized deleted residuals are then centered by subtracting the estimated median. For the ith observation we let $d_{c,i}$ denote the centered studentized deleted residual and we let $D_c = \{d_{c,1}, \ldots, d_{c,n}\}$. Next we compute $\hat{F}_h(d_{c,i};\ D_c)$, which is the smoothed c.d.f. evaluated at $d_{c,i}$. We then compute $t_i = T^{-1}_{n-p-1}(\hat{F}_h(d_{c,i};\ D_c))$ using the inverse of the c.d.f. of the t distribution with $n - p - 1$ degrees of freedom. The adaptive weights are then computed using $w_i = t_i / d_{c,i}$ for $i = 1, \ldots, n$. If the errors are normally distributed, then the smoothed and centered c.d.f. should approximate the c.d.f. of a t distribution with $n - p - 1$ degrees of freedom and the weights should approximate one. However, if an outlier produces a studentized deleted residual that is quite large, then the weight will be small for that observation.

For the two-sided adaptive test of significance these weights are then used in a WLS procedure to compute the test statistics for testing $H_0 : \beta_{ADJ} = 0$ versus $H_a : \beta_{ADJ} \neq 0$, which is equivalent to testing $H_0 : \beta_A = c$ versus $H_a : \beta_A \neq c$. To perform the adaptive test we permute the rows of X_A many times, compute the test statistic from each permutation, and then compute the p-value.

To compute the confidence interval limits we perform a one-sided test of $H_0 : \beta_{ADJ} = 0$ versus $H_a : \beta_{ADJ} < 0$ and of $H_0 : \beta_{ADJ} = 0$ versus $H_a : \beta_{ADJ} > 0$. We weight the observations as we do in a test of significance and compute the test statistic $t = \hat{\beta}_A/\text{s.e.}\{\hat{\beta}_A\}$ based on the weighted model. To compute the estimate of the lower limit of a $100(1-\alpha)\%$ confidence interval we use an iterative method, with L_i as the estimate at the ith iteration. To adjust the lower limit at the ith iteration we use the adaptive test procedure to compute the unpermuted t test statistic for $H_0 : \beta_A = L_i$, which is denoted by t^*. Next, one permutation of X_A is performed and the permutation test statistic t^{perm} is computed. These test statistics are used to update the estimate of the lower limit using a steplength value of k_i in the updating formula

$$L_{i+1} = \begin{cases} L_i + k_i\left(\alpha/2\right)/i & \text{if } t^{perm} < t*, \\ L_i - k_i\left(1-\alpha/2\right)/i & \text{if } t^{perm} \geq t^*. \end{cases}$$

For $\alpha = .05$ Garthwaite (1996) recommended using $k_i = 17.46(\hat{B}_A - L_i)$, where $\hat{B}_A$ is the adaptive estimate of B_A. The adaptive estimate is obtained by using the complete model to determine the weights, which are then used to obtain the WLS estimate. The adaptive estimation procedure will be described fully in Chapter 7.

We start the iteration procedure to calculate a 95% confidence interval by using the limits from the OLS model as the initial estimates. Following the recommendation of Garthwaite (1996), we did not start the iterations with $i = 1$ but instead with

$i = \lfloor .3(2-\alpha/2)/(\alpha/2) \rfloor$, which for $\alpha = .05$ gives $i = 23$. The iterative method was used to adjust the values of L_i, and successive blocks of 1000 iterations were used until the convergence criteria were satisfied. The convergence criteria used were the same as those described in section 6.2. This stopping rule required at least 6999 iterations but seldom more than 12,999 iterations.

This same general method was used to compute the upper limit. The updating formula is

$$U_{i+1} = \begin{cases} U_i - k_i\left(\alpha/2\right)/i & \text{if } t^{perm} > t*, \\ U_i + k_i\left(1-\alpha/2\right)/i & \text{if } t^{perm} \le t^*, \end{cases}$$

with a steplength value of $k_i = 17.46(U_i - \hat{B}_A)$. The iterations were started at $i = 23$ and the convergence criteria for the upper limit were similar to those for the lower limit.

It should be noted that we do not need to recompute the weights in order to compute t^{perm} since the weights were based on the reduced model, which is not involved in the permutation. However, each time we change one of the limits the values in y_{ADJ} will change so the weights will need to be recomputed. Since each limit is adjusted approximately 10,000 times and since smoothing the c.d.f. of the residuals requires many calculations when n is large, the adaptive confidence interval procedure becomes computationally intensive for large n.

In the remaining sections of this chapter we will compute adaptive confidence intervals for some important special cases. The adaptive confidence interval for the difference between two means will be described in section 6.5 and that for the slope in a simple linear regression will be described in section 6.6. An adaptive confidence interval for one slope in a multiple linear regression model will be described in section 6.7.

6.5 An Adaptive Confidence Interval for the Difference between Two Means

An adaptive confidence interval for the difference between two population means can be obtained by using the general procedure described in section 6.4. The model $Y = X_R\beta_R + X_A\beta_A + \varepsilon$ can be used to compute this confidence interval by setting X_R equal to 1, which is an $n \times 1$ vector of ones, and by using indicator variables in the $n \times 1$ vector X_A. The model then becomes $Y = 1\beta_R + X_A\beta_A + \varepsilon$, where the ith element in X_A is given by

$$X_{A,i} = \begin{cases} 1 & \text{if the } i\text{th observation is from population 2,} \\ 0 & \text{if the } i\text{th observation is from population 1.} \end{cases}$$

With this parameterization, β_A equals the mean of the second population minus the mean of the first population.

In our search for the upper and lower limits we will adjust Y to obtain Y_{ADJ} and then use the reduced model $Y_{ADJ} = 1\beta_R + \varepsilon$ to determine the appropriate weights. After weighting, the t^* test statistic is obtained from the unpermuted data while the t^{perm} statistic is obtained by permuting the rows of X_A, which is equivalent to shuffling the group membership.

6.5.1 An Example

Anderson et al. (1984) published results on an experiment designed to determine the effects of certain high fiber foods on cholesterol concentrations. In their experiment 20 hypercholesterolemic men were randomly assigned to a bean or an oat-bran supplemented diet. After 21 days the 10 men who were assigned to the bean diet had final triglyceride levels (mg/dL) of {175, 173, 110, 116, 136, 209, 351, 176, 341, 476}. The 10 men who were assigned to the oat-bran diet had final triglyceride levels (mg/dL) of {159, 155, 144, 101, 196, 150, 195, 270, 456, 523}. The relationship between the bean triglyceride levels and the oat-bran triglyceride levels can be seen more clearly in the dot plot shown in Figure 6.4. The dot plot shows that both groups have highly variable triglyceride levels and that the distributions are skewed to the right.

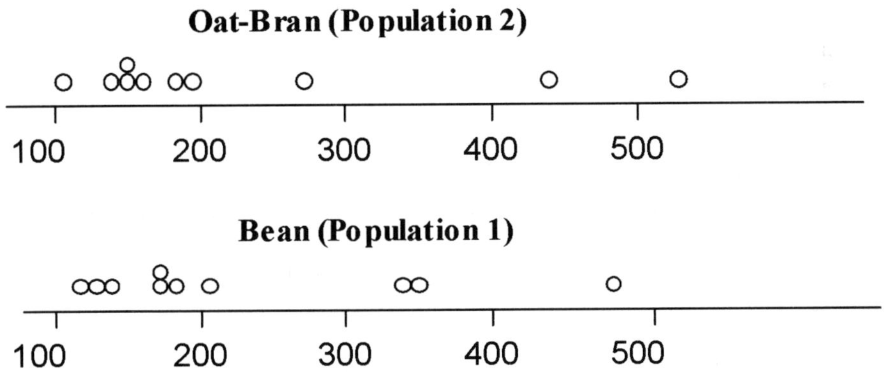

Figure 6.4. *Final triglyceride levels of ten subjects on a diet supplemented by beans and ten subjects on a diet supplemented by oat bran.*

In order to compute an adaptive confidence interval for the difference between the population means, we used the model $Y = 1\beta_R + X_A\beta_A + \varepsilon$, where the ith element in X_A is given by

$$X_{A,i} = \begin{cases} 1 & \text{if the } i\text{th subject has been assigned to an oat-bran diet,} \\ 0 & \text{if the } i\text{th subject has been assigned to a bean diet.} \end{cases}$$

The traditional method gives a 95% confidence interval of $[-115.7,\ 132.9]$. These traditional limits were used as starting values of L_i and U_i in the iterative adaptive procedure. The adaptive procedure used the iterative process until the convergence criteria were satisfied to produce a lower limit of -85.7 and an upper limit of 86.0. The lower limit converged when $i = 7999$ and the upper limit converged when $i = 8999$.

To illustrate how quickly the iterative procedure converged, the iterative history of the lower and upper limits is shown in Figure 6.5. Although the search procedure is somewhat erratic at first, there is little meaningful change in these limits after 3000 iterations.

Since, for this example, the adaptive confidence interval is much narrower than the traditional confidence interval, the adaptive method appears to be more effective than the traditional method. The adaptive procedure was able to make the interval narrower in

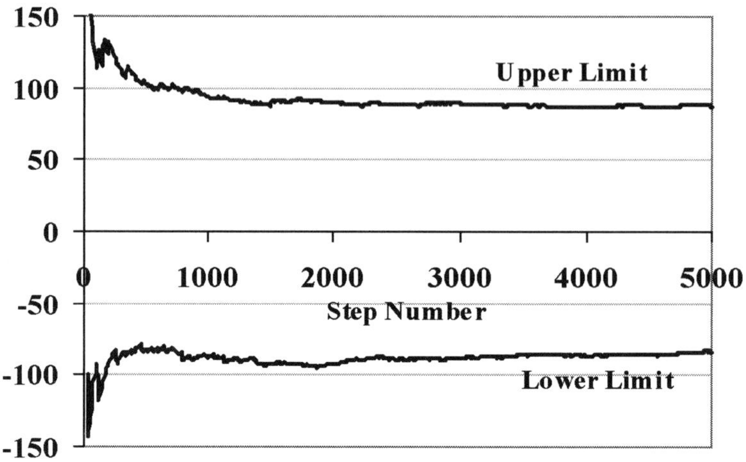

Figure 6.5. *The iterative history of the upper and lower confidence limits for the triglyceride data set.*

this example by downweighting the few subjects with unusually large triglyceride values. However, the results from this example may not be representative of what would happen if we used adaptive confidence intervals for other data sets. Consequently, to fully evaluate the performance of the adaptive method we will use the simulation studies described in the next section.

It is not difficult to compute these adaptive confidence intervals. The SAS macro "ciawls" listed in Appendix A can be used to compute the 95% confidence interval for a single parameter in a multiple regression model. If we had the data in a SAS data set named "fiber" with the dependent variable named "tri" and the group indicator named "diet," we could obtain the 95% confidence interval by including the SAS macro and by invoking the "ciawls" macro in the SAS program with the macro call

%ciawls(dataset=fiber, y=tri, xr=, xa=diet);

Note that the columns in the X_R matrix do not need to be specified in this example because SAS will always include an intercept in the model. With this macro, the number of variables in the reduced model does not need to be specified; it will be calculated by the macro as the number of variables listed in the "xr" parameter. Unlike the FORTRAN programs used in the simulation studies, this SAS macro does not use the convergence criteria. Instead, it uses 8000 iterations, by default, or the number of iterations specified by the "maxiter" parameter. This macro may run slowly on some computers because the iterative process requires many calculations in the smoothing of the c.d.f. of the deleted residuals.

6.5.2 Coverage Probabilities and Average Widths

In order to evaluate the effectiveness of the adaptive confidence intervals for the difference between two population means, we need to determine if they are valid. A simulation study was used to determine if the adaptive 95% confidence intervals really did include the population difference for approximately 95% of the data sets. For this simulation study we used the same nine error distributions as in the simulations in Chapter 4. Let n_1 equal the number of observations from population 1 and n_2 the number of observations from population 2. Several balanced sample sizes were used in this study, ranging from $n_1 = n_2 = 5$ to $n_1 = n_2 = 50$. The simulation study also used the unbalanced sample size of $n_1 = 12,\ n_2 = 48$. For each sample size and error distribution 1000 data sets were generated.

The empirical coverage probabilities are given in Table 6.3. Since 1000 data sets were used the standard error was approximately 0.7%. These results indicate that both the adaptive and traditional confidence intervals have coverage probabilities that approximate their nominal values. Consequently, we can fairly evaluate the relative performance of these methods by comparing their average widths.

Table 6.3. *The empirical coverage probabilities in percent for the adaptive and the traditional* 95% *confidence interval for the differences between the means of two populations.*

	Distributions								
Skewness	0	0	0	1	1	1	2	2	2
Kurtosis	1.8	3.0	6.0	4.2	5.4	8.4	11.4	12.6	15.6
				$n_1 = 5,\ n_2 = 5$					
Adaptive	94.8	95.7	94.6	95.5	95.0	95.3	94.9	93.5	95.2
Traditional	94.9	96.3	95.6	96.1	95.7	96.6	95.7	95.2	96.2
				$n_1 = 10,\ n_2 = 10$					
Adaptive	94.8	94.9	95.3	94.2	94.4	94.4	94.9	95.2	94.5
Traditional	94.8	95.4	94.9	94.4	94.3	95.1	95.7	96.0	94.4
				$n_1 = 20,\ n_2 = 20$					
Adaptive	95.5	95.4	94.2	94.6	94.8	94.7	94.0	95.1	95.1
Traditional	95.3	96.0	94.9	95.8	95.2	94.7	95.4	95.9	96.2
				$n_1 = 50,\ n_2 = 50$					
Adaptive	94.9	96.1	95.8	95.7	96.3	93.7	94.4	94.9	94.7
Traditional	94.8	95.5	95.9	94.8	95.3	94.2	93.8	95.0	94.9
				$n_1 = 12,\ n_2 = 48$					
Adaptive	93.5	95.6	94.1	94.7	95.4	94.8	95.3	96.2	96.3
Traditional	94.5	95.5	93.9	95.2	95.0	94.5	95.0	95.8	95.1

The average widths of the adaptive and traditional confidence intervals and the relative efficiencies of the adaptive to the traditional confidence interval are presented in Table 6.4. These results are similar to those found in O'Gorman (2001c) for an earlier version of the adaptive confidence interval. A scatterplot of the adaptive and traditional confidence intervals in O'Gorman (2001c) demonstrated that it was appropriate to use the average widths of the confidence intervals to compare the performance of the methods. The results in Table 6.4 show that the adaptive confidence intervals have approximately the same average width as the traditional confidence intervals for small data sets with $n_1 = n_2 = 5$ observations. The results also show that, if $n_1 \geq 10$ and $n_2 \geq 10$, the adaptive intervals are generally narrower

Table 6.4. *The average widths of the adaptive and traditional 95% confidence intervals for the differences between the means of two populations, and the relative efficiencies of the adaptive confidence interval to the traditional confidence interval.*

	Distributions								
Skewness	0	0	0	1	1	1	2	2	2
Kurtosis	1.8	3.0	6.0	4.2	5.4	8.4	11.4	12.6	15.6
	$n_1 = 5, n_2 = 5$								
Adaptive	2.83	2.81	2.73	2.80	2.75	2.71	2.65	2.60	2.61
Traditional	2.84	2.83	2.75	2.83	2.78	2.74	2.69	2.64	2.64
Relative Eff.	1.01	1.01	1.02	1.02	1.02	1.02	1.03	1.03	1.02
	$n_1 = 10, n_2 = 10$								
Adaptive	1.85	2.12	1.79	1.73	1.74	1.71	1.54	1.55	1.58
Traditional	1.86	2.10	1.82	1.82	1.83	1.81	1.78	1.82	1.79
Relative Eff.	1.01	0.98	1.04	1.11	1.10	1.11	1.33	1.38	1.28
	$n_1 = 20, n_2 = 20$								
Adaptive	1.19	1.89	1.19	1.12	1.16	1.14	0.97	1.00	1.02
Traditional	1.28	1.87	1.27	1.26	1.27	1.24	1.24	1.25	1.25
Relative Eff.	1.14	0.97	1.12	1.28	1.19	1.20	1.63	1.57	1.50
	$n_1 = 50, n_2 = 50$								
Adaptive	0.66	0.80	0.73	0.66	0.70	0.70	0.56	0.59	0.62
Traditional	0.79	0.79	0.79	0.79	0.79	0.79	0.79	0.78	0.78
Relative Eff.	1.43	0.98	1.15	1.43	1.27	1.27	1.98	1.79	1.62
	$n_1 = 12, n_2 = 48$								
Adaptive	1.16	1.31	1.20	1.11	1.16	1.16	0.95	0.99	1.03
Traditional	1.29	1.29	1.28	1.28	1.28	1.29	1.27	1.27	1.27
Relative Eff.	1.23	0.97	1.14	1.32	1.21	1.24	1.78	1.64	1.52

than the traditional intervals for nonnormal error terms. If the errors are approximately normal, then the adaptive intervals are only slightly wider, on average, than the traditional intervals.

6.6 An Adaptive Confidence Interval for the Slope in a Linear Regression

6.6.1 An Example

As part of a research effort to determine the nutritional status of a population of the elderly in the Dominican Republic, Simon (2001) collected health-related information on 67 males between the ages of 55 and 79. These data included height, weight, and systolic blood pressure. Based on these measurements, we have computed the body mass index (BMI) and created the scatterplot shown in Figure 6.6. The data in the scatterplot suggest, but do not clearly demonstrate, that the average systolic blood pressure may increase with an increase in BMI.

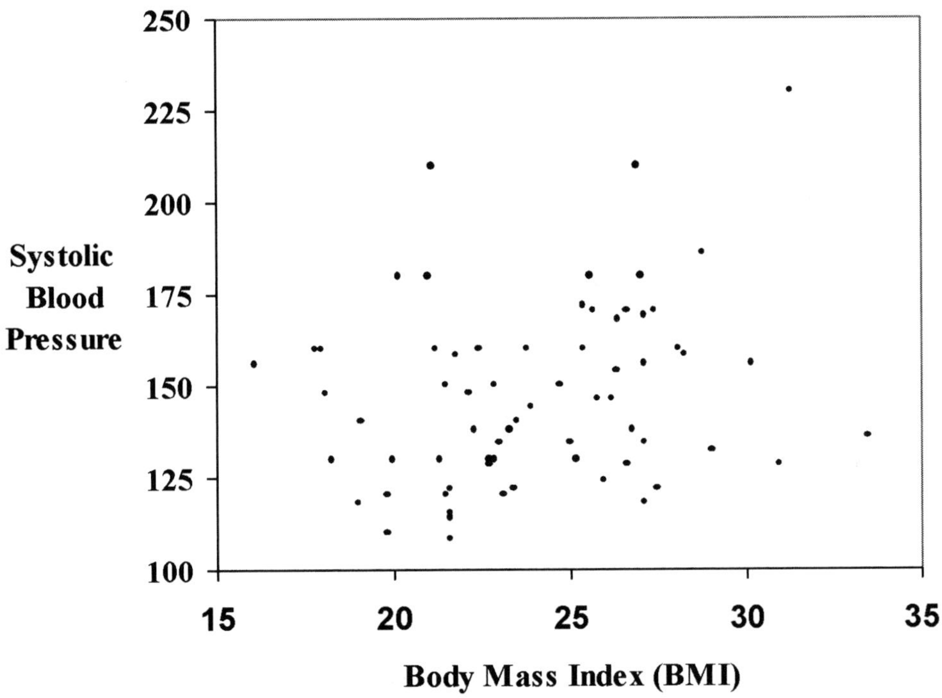

Figure 6.6. *Scatterplot of systolic blood pressure and BMI for* $n = 67$ *males between the ages of* 55 *and* 79.

In order to quantify the increase in systolic blood pressure with BMI, we used a simple linear regression model $Y = \beta_0 + \beta_1 X + \varepsilon$, where Y is the systolic blood pressure and X is the BMI. Our objective was to obtain a 95% confidence interval for β_1. The traditional analysis gave a 95% confidence interval of [0.005, 3.319]. The adaptive confidence interval procedure described in section 6.4 gave a 95% confidence interval of [0.030, 2.805]. The adaptive method used 11,999 iterations for the lower limit and 7999 iterations for the upper limit before the convergence criteria were satisfied. Since the width of the adaptive interval is 2.775 and the width of the traditional interval is 3.314, the adaptive method was effective in reducing the width for this confidence interval. Of course, we do not know which interval, if any, actually includes β_1. Consequently, we cannot be too confident that the adaptive method is superior to the traditional method unless we look carefully at the coverage probabilities and the average widths of these intervals.

We could use the same SAS macro as we used in the previous section to perform the calculations for the 95% confidence interval. For this example we will assume that the SAS data set is named "health," the dependent variable is named "sbp," and the independent variable is named "bmi." The SAS macro can be invoked by using the statement

```
%ciawls(dataset=health, y=sbp, xr=, xa=bmi, maxiter=4000);
```

In this example the matrix X_R consists of a column of ones so, by not specifying anything in the "xr" list, SAS will include an intercept term. The SAS macro does not use the convergence criteria. In this example we have used a maximum number of iterations of 4000 for the upper and lower limits, which should be sufficient to give a reasonably accurate answer. If more precision is needed, the "maxiter" parameter could be set to 10,000 or 20,000 iterations with a corresponding increase in computing time.

6.6.2 Coverage Probabilities and Average Widths

A Monte Carlo simulation study was performed in order to evaluate the coverage probabilities and average widths of the traditional and adaptive confidence intervals for the slope in a simple linear regression. These simulations used sample sizes of $n = 10, 20$, and 50 with the simple linear regression model $Y = \beta_0 + \beta_1 X + \varepsilon$. The random errors in these models were generated from one of the nine generalized lambda distributions. For each sample size and error distribution 1000 data sets were generated. For most of the simulations we computed the 95% confidence intervals, but we did include some simulation results for 90% and 99% confidence intervals. With 1000 data sets used to produce the empirical coverage probabilities, the standard errors for the 90%, 95%, and 99% confidence intervals are 1.0%, 0.7%, and 0.3%, respectively. The empirical coverage probabilities in Table 6.5 clearly show that both the traditional and adaptive confidence intervals have coverage probabilities that are near their nominal values. Because both methods are robust for validity we can proceed to compare the average widths of the confidence intervals.

Table 6.5. *The empirical coverage probabilities in percent of the adaptive and traditional confidence intervals for the slope in a simple linear regression.*

	Distributions								
Skewness	0	0	0	1	1	1	2	2	2
Kurtosis	1.8	3.0	6.0	4.2	5.4	8.4	11.4	12.6	15.6
	95% Confidence Interval with $n = 10$								
Adaptive	95.3	95.2	94.7	96.0	95.9	94.6	95.3	95.0	95.0
Traditional	95.7	95.2	94.5	96.3	95.4	94.3	95.7	94.9	95.5
	95% Confidence Interval with $n = 20$								
Adaptive	95.2	94.7	96.3	94.4	94.5	95.2	94.1	94.9	93.5
Traditional	95.3	95.0	95.8	95.3	94.6	95.0	93.9	94.8	94.1
	95% Confidence Interval with $n = 50$								
Adaptive	94.4	94.4	94.2	93.9	95.2	95.8	95.6	95.0	93.8
Traditional	94.1	95.2	94.2	94.9	95.3	95.5	95.6	93.8	93.8
	90% Confidence Interval with $n = 20$								
Adaptive	90.6	90.1	89.6	90.7	90.3	90.4	89.1	91.7	88.6
Traditional	90.8	90.1	91.4	89.7	91.1	89.9	89.9	92.4	89.7
	99% Confidence Interval with $n = 20$								
Adaptive	99.2	98.9	99.2	98.6	99.2	99.1	99.3	99.5	99.2
Traditional	99.1	98.7	98.9	98.5	99.5	98.9	99.4	99.0	99.2

The average widths over the 1000 data sets used in each simulation are displayed in Table 6.6 along with the relative efficiencies of the adaptive confidence interval to the traditional confidence interval. The results show that the width of the adaptive interval approximates the width of the traditional interval for $n = 10$, which implies that there is no good reason to use the adaptive confidence interval for such small samples. However, for the larger samples with $n = 20$ and $n = 50$, the adaptive confidence intervals were generally shorter, on average, than the traditional intervals for nonnormal error distributions. We also note that when the errors are generated by a normal distribution the adaptive confidence intervals are only slightly wider than the traditional confidence intervals. The relative efficiencies of the adaptive confidence interval to the traditional confidence interval for 90% confidence intervals with $n = 20$ are close to the relative efficiencies for the 95% confidence intervals. However, the relative efficiencies with nonnormal distributions for the 99% confidence intervals are not as large as those for the 95% confidence intervals, indicating that there is less of an advantage to using adaptive methods with 99% confidence intervals compared to 95% confidence intervals. With 99% confidence intervals the reason that the adaptive methods lose some of their advantage over traditional methods is that the weights are not as appropriate because they are based on values of L_i and U_i that are often farther from β_A.

Table 6.6. *The average widths of the adaptive and traditional confidence intervals for the slope in a simple linear regression, and the relative efficiencies of the adaptive confidence interval to the traditional confidence interval.*

	Distributions								
Skewness	0	0	0	1	1	1	2	2	2
Kurtosis	1.8	3.0	6.0	4.2	5.4	8.4	11.4	12.6	15.6
	95% Confidence Interval with $n = 10$								
Adaptive	1.74	1.70	1.61	1.66	1.63	1.54	1.56	1.54	1.50
Traditional	1.68	1.64	1.58	1.64	1.60	1.53	1.59	1.54	1.53
Relative Eff.	0.93	0.93	0.96	0.97	0.96	0.99	1.04	1.00	1.04
	95% Confidence Interval with $n = 20$								
Adaptive	1.00	1.00	0.97	0.92	0.94	0.92	0.82	0.83	0.83
Traditional	1.00	0.98	0.99	0.98	0.98	0.97	0.95	0.96	0.94
Relative Eff.	1.00	0.95	1.03	1.12	1.10	1.12	1.35	1.34	1.29
	95% Confidence Interval with $n = 50$								
Adaptive	0.53	0.58	0.54	0.50	0.52	0.52	0.43	0.44	0.46
Traditional	0.58	0.58	0.58	0.58	0.58	0.57	0.57	0.56	0.56
Relative Eff.	1.20	0.97	1.12	1.31	1.22	1.22	1.73	1.62	1.48
	90% Confidence Interval with $n = 20$								
Adaptive	0.82	0.85	0.77	0.76	0.77	0.75	0.66	0.67	0.69
Traditional	0.83	0.82	0.79	0.80	0.82	0.80	0.80	0.79	0.78
Relative Eff.	1.02	0.94	1.05	1.12	1.13	1.14	1.49	1.39	1.28
	99% Confidence Interval with $n = 20$								
Adaptive	1.39	1.40	1.33	1.31	1.35	1.31	1.22	1.23	1.23
Traditional	1.36	1.36	1.33	1.34	1.36	1.34	1.31	1.32	1.31
Relative Eff.	0.96	0.95	1.00	1.04	1.02	1.06	1.15	1.16	1.13

6.7 A Confidence Interval for One Slope in a Multiple Regression

6.7.1 An Example

The effectiveness of the adaptive confidence interval procedure for estimating a single parameter in a multiple regression model can be illustrated by the analysis of the Scottish Hills Race data in Table 6.7. These data, which were provided by the Scottish Hills Runners

Table 6.7. *Scottish Hills Race data of* 1984.

Race	Name	Time (s)	Climb (ft.)	Distance (mi.)
1	Greenmantle New Year Dash	965	650	2.5
2	Carnethy	2901	2500	6
3	Craig Dunain	2019	900	6
4	Ben Rha	2736	800	7.5
5	Ben Lomond	3736	3070	8
6	Goatfell	4393	2866	8
7	Bens of Jura	12277	7500	16
8	Cairnpapple	2182	800	6
9	Scolty	1785	800	5
10	Traprain Law	2385	650	6
11	Lairig Ghru	11560	2100	28
12	Dollar	2583	2000	5
13	Lomonds of Fife	3900	2200	9.5
14	Cairn Table	2648	500	6
15	Eildon Two	1616	1500	4.5
16	Cairngorm	4335	3000	10
17	Seven Hills of Edinburgh	5905	2200	14
18	Knock Hill	4719	350	3
19	Black Hill	1045	1000	4.5
20	Creag Beag	1954	600	5.5
21	Kildoon	957	300	3
22	Meall Ant-Suiche	1674	1500	3.5
23	Half Ben Nevis	2859	2200	6
24	Cow Hill	1076	900	2
25	North Berwick Law	1121	600	3
26	Creag Dubh	1573	2000	4
27	Burnswark	2066	800	6
28	Largo	1714	950	5
29	Criffel	3030	1750	6.5
30	Achmony	1257	500	5
31	Ben Nevis	5135	4400	10
32	Knockfarrel	1943	600	6
33	Two Breweries Fell	10215	5200	18
34	Cockleroi	1686	850	4.5
35	Moffat Chase	9590	5000	20

Association on the races run in 1984, were first published by Atkinson (1986). The data include the time to complete the race, the climb, and the distance. We will use the time as the dependent variable, with the climb and the distance being the two independent variables.

If we use OLS methods, we obtain the fitted equation as

$$Y = -539.48 + 0.66289X_1 + 373.07X_2,$$

where Y is the time, X_1 is the climb, and X_2 is the distance. Since the time to complete a race is clearly a function of the distance, tests of significance are not appropriate. Instead, we will compute the traditional and adaptive confidence intervals for β_2. Using the traditional approach we obtain a 95% confidence interval of $[299.69, 446.45]$. Using the adaptive confidence interval procedure we obtain a 95% confidence interval of $[358.96, 427.03]$, which is much narrower than the traditional interval.

The large difference between the traditional confidence interval and the adaptive confidence interval can be explained, in part, by the presence of two outliers in the data. Observation 7, the "Bens of Jura" race, had the longest time and the greatest climb. Observation 18, the "Knock Hill" race, seemed to take a long time considering that it was only a 3-mile race.

These data are interesting because two years after the publication of the original data Atkinson (1988) reported that the time for race 18 should have been 1119 seconds instead of the 4719 seconds originally reported. If we use the corrected value for that race in the analysis, we obtain a traditional 95% confidence interval of $[336.73, 424.73]$ and an adaptive 95% confidence interval of $[359.24, 428.74]$, so even with the corrected data the adaptive method appears to more accurately estimate the relationship between the distance and the race time. The Scottish Hills Race data have also been analyzed by Chatterjee, Hadi, and Price (2000, Chap. 4) and by Ryan (1997, Chap. 15).

The SAS macro "ciawls" can be used to compute the 95% confidence interval for β_2. If we create a SAS data set named "racedata," we can invoke the macro with the following statement:

```
%ciawls(dataset=racedata, y=time, xr=climb, xa=distance,
        seed=13721, maxiter=6000);
```

In this invocation we set a starting seed for the random number generator and use 6000 iterations for the lower and upper limits.

6.7.2 Coverage Probabilities and Average Widths

A simulation study was performed in order to determine if the adaptive confidence interval for one parameter in a linear model was effective in reducing the average width of the interval for nonnormal error distributions. Our objective with this study was to evaluate the performance of a 95% confidence interval for β_2 in the linear model $y = \beta_0+\beta_1x_1+\beta_2x_2+\varepsilon$. The independent variables were generated from a bivariate normal distribution with zero means, unit variances, and a correlation of $\rho = 0.8$. Thus, the independent variables are highly correlated. As we have done in previous simulations, we generate the error terms from one of nine generalized lambda distributions.

Table 6.8. *The empirical coverage probabilities in percent for the confidence interval for one slope in a multiple linear regression.*

	Distributions								
Skewness	0	0	0	1	1	1	2	2	2
Kurtosis	1.8	3.0	6.0	4.2	5.4	8.4	11.4	12.6	15.6
					$n = 10$				
Adaptive	94.7	95.2	94.1	96.1	95.7	94.7	94.8	95.9	95.8
Traditional	94.7	95.2	94.7	95.3	95.6	94.7	94.3	96.4	96.3
					$n = 20$				
Adaptive	93.6	95.8	96.2	95.8	96.6	94.8	95.3	95.5	94.4
Traditional	93.9	96.2	96.2	95.2	96.1	94.2	95.5	95.0	94.7
					$n = 50$				
Adaptive	95.3	96.1	95.4	94.6	96.3	95.3	94.9	96.4	95.3
Traditional	95.6	95.6	95.3	95.0	95.8	94.4	95.2	95.3	95.9

The coverage probabilities for $n = 10$, 20, and 50 are given in Table 6.8 from simulations that used 1000 data sets. As we expected, the simulation studies demonstrated that the traditional confidence intervals and the adaptive confidence intervals maintain their coverage probabilities near the nominal level.

The average widths of the adaptive and traditional confidence intervals are given in Table 6.9 for these simulations. These average widths, and the relative efficiencies determined from them, show that the adaptive confidence intervals are narrower, on average, than the traditional confidence intervals for nonnormal error distributions if $n \geq 20$. The simulations also show that there is no advantage of the adaptive confidence interval over the traditional confidence interval if $n = 10$.

It is instructive to compare the results for a single slope in a multiple regression model to the results for a simple linear regression model. We note that the results in Table 6.9 for multiple linear regression are quite similar to those in Table 6.6 for simple linear regression. However, the relative efficiencies in Table 6.9 tend to be slightly smaller than the corresponding relative efficiencies in Table 6.6.

6.8 Concluding Remarks

We used the relationship between tests and confidence intervals to construct the adaptive confidence intervals in the hope that the good properties that we have observed with the adaptive tests would carry over to the adaptive confidence intervals. Indeed, by comparing the relative efficiencies of the confidence intervals in this chapter to the relative performance of the corresponding adaptive tests in the previous chapter, we can see that the properties of the adaptive confidence intervals parallel the properties of the adaptive tests. The adaptive tests and the adaptive confidence intervals are robust for validity and for efficiency.

In the adaptive test for paired data we assumed that the observations were symmetrically distributed. Symmetry is implied by the paired data model if the distribution of one member of the pair is shifted relative to the distribution of the other. However, with paired

Table 6.9. *The average widths of the adaptive and traditional confidence intervals for one slope in a multiple linear regression, and the relative efficiencies of the adaptive confidence interval to the traditional confidence interval.*

	Distributions								
Skewness	0	0	0	1	1	1	2	2	2
Kurtosis	1.8	3.0	6.0	4.2	5.4	8.4	11.4	12.6	15.6
					$n = 10$				
Adaptive	3.13	3.05	3.01	3.08	3.01	3.03	2.85	2.81	2.82
Traditional	2.99	2.93	2.96	3.01	2.92	2.98	2.88	2.81	2.82
Relative Eff.	0.91	0.92	0.96	0.96	0.94	0.97	1.02	0.99	1.02
					$n = 20$				
Adaptive	1.72	1.76	1.66	1.63	1.63	1.61	1.46	1.47	1.49
Traditional	1.71	1.72	1.70	1.69	1.67	1.67	1.64	1.66	1.63
Relative Eff.	0.99	0.95	1.04	1.09	1.06	1.07	1.26	1.29	1.19
					$n = 50$				
Adaptive	0.91	0.99	0.93	0.86	0.89	0.88	0.74	0.77	0.79
Traditional	0.98	0.98	0.98	0.97	0.98	0.98	0.96	0.98	0.95
Relative Eff.	1.16	0.97	1.11	1.30	1.20	1.22	1.70	1.61	1.47

data the shift model may not always hold, so the confidence interval for the mean of a symmetric distribution may not always be appropriate. In addition, the confidence interval for the mean of a symmetric distribution is often not appropriate for one-sample data that are not derived from paired data.

The key properties of the adaptive confidence intervals can be summarized as follows:

1. Adaptive confidence intervals have coverage probabilities that are near the nominal level.

2. For $n \leq 10$ adaptive confidence intervals do not perform much better than traditional confidence intervals.

3. For $n \geq 20$ adaptive confidence intervals are narrower, on average, than traditional confidence intervals if the error distributions are nonnormal.

4. For normal error distributions, adaptive methods are only slightly wider than traditional confidence intervals.

The adaptive confidence intervals that we constructed used an iterative method suggested by Garthwaite (1996). However, this method involves a considerable amount of calculation, primarily because the deleted residuals need to be smoothed at each iteration. This is the main reason why the SAS macros for confidence intervals do not execute quickly. Further research in this area may lead to more efficient methods of calculation.

Chapter 7
Adaptive Estimation

7.1 An Introduction to Adaptive Estimation

Adaptive estimation and robust estimation are closely related. The objective of robust estimation is to improve the accuracy of the estimate by reducing the influence of outliers. Adaptive estimation can be considered as a type of robust estimation because, when some observations are adaptively downweighted, the influence of outliers is greatly reduced. Both robust and adaptive estimators can be effective in achieving a relatively low mean squared error (m.s.e.) for a variety of nonnormal distributions of errors.

Although robust estimation methods and adaptive estimation methods have much in common, there are important differences between them. Robust estimators are usually designed to reduce the influence of observations that are judged to be extreme, even if the observations appear to roughly follow a normal distribution. In contrast, adaptive estimators take a preliminary look at the data before deciding if any reduction in influence is required. We will illustrate the difference between these approaches by comparing a robust estimator to an adaptive estimator of the mean of a symmetric distribution.

7.1.1 Adaptive and Robust Estimation of the Mean of a Symmetric Distribution

Suppose we have n observations from a symmetric distribution with an unknown mean μ. One popular robust estimator of μ is the 10% trimmed mean, which is obtained by removing the largest and smallest 10% of the observations before the average is computed. When the 10% trimmed mean was used with real data sets, Stigler (1977) found that it was superior to several other more complex estimators. The trimmed mean is nonadaptive because it trims observations even if trimming is not necessary. For example, if we have $n = 50$ observations from a normal distribution, the 10% trimmed mean will trim the largest and the smallest five observations, even though this trimming will slightly increase the variance of the estimator (see Bickel and Doksum (1977, p. 371)). If the observations are from a contaminated normal distribution, the 10% trimmed mean may be very effective, provided the percentage of observations from the contaminating distribution is less than 20%, but the trimming may not be effective if the contaminated normal distribution has a large percentage of outliers. An extensive study of many of these nonadaptive estimators was published by Andrews et al. (1972). Reed and Stark (1994) also evaluated several adaptive estimators.

Over the last 20 years several adaptive estimators of the mean of a symmetric distribution have been proposed. The adaptive estimator described by Hogg and Lenth (1984) uses a selection statistic to determine if trimming is necessary. If no trimming appears to be required, then the estimator is simply the mean; if some trimming is required, then the estimator is the 25% trimmed mean; if a great deal of trimming is necessary, then the estimator is the median. The selection statistic used in this adaptive procedure is

$$Q = \frac{\bar{U}_{.2} - \bar{L}_{.2}}{\bar{U}_{.5} - \bar{L}_{.5}},$$

where $\bar{U}_p$ is the average of the largest np order statistics and $\bar{L}_p$ is the average of the smallest np order statistics. In the adaptive estimation procedure of Hogg and Lenth (1984), the mean is selected as the estimator if the empirical distribution has short tails with $Q < 1.81$, the 25% trimmed mean is selected if the empirical distribution has moderate tails with $1.81 \leq Q \leq 1.87$, and the median is selected if the empirical distribution has tails that are so long that $Q > 1.87$. This estimation procedure, which is called Hogg's adaptive estimator, avoids unnecessary trimming of extreme values while making the estimator reasonably robust to the presence of a high percentage of outliers. In section 7.2 we will compare Hogg's adaptive estimator to the 10% trimmed mean and to an adaptive estimator based on a WLS approach. More recently, Hill and Padmanabhan (1991) proposed and evaluated the performance of an adaptive estimator.

7.1.2 Robust Regression

In addition to adaptively estimating the mean, we would like to adaptively estimate the parameters in a linear regression model. Before developing an adaptive version of a robust regression estimator, it will be helpful to describe several nonadaptive robust regression methods.

One class of robust estimators includes the high breakdown point estimators that attempt to robustify the regression estimators by ignoring a portion of the data. One example of a high breakdown estimator is the LMS estimator, which produces estimates of the parameters that minimize the median of the squares of the ordinary residuals. The LMS estimator, which was introduced by Rousseeuw (1984), has a certain appeal as a robust estimator because it ignores a large percentage of observations. If we are attempting to obtain an estimate for the slope in a simple linear regression for the data shown in the scatterplot in Figure 7.1, the LMS estimate is the slope of the line in the center of the narrowest strip that contains 50% of the data. In this example the LMS estimator completely ignored the outliers to obtain a very reasonable estimate of the slope. The problem with the LMS estimator is that it does not always produce reasonable and stable estimates.

One serious problem with the LMS estimator is that, for some data sets, it can be extremely sensitive to small changes in the data. This sensitivity was pointed out by Hettmansperger and Sheather (1992), who cautioned that the estimator should not be used blindly. The reasons for this instability can be demonstrated by the scatterplot shown in

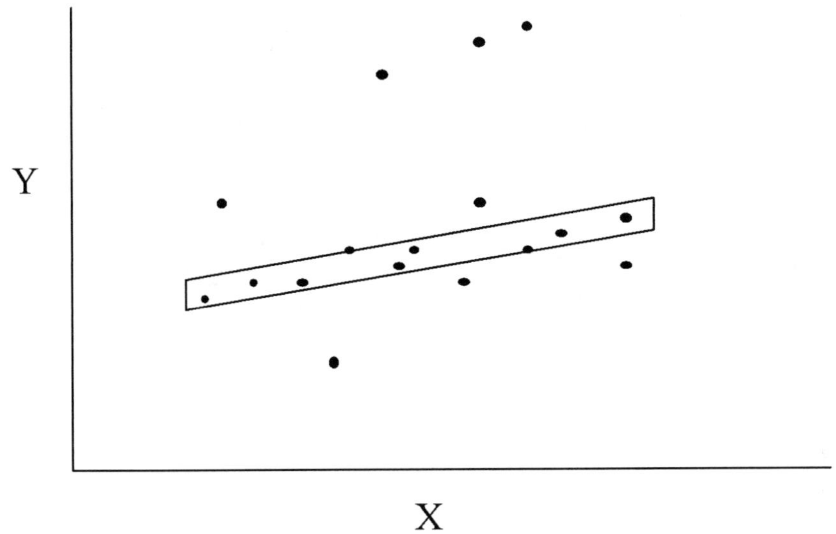

Figure 7.1. *A scatterplot that illustrates the LMS estimator.*

Figure 7.2, which displays 12 data points represented by solid circles and one other observation, indicated by an open circle, that is at Position A or Position B. Since there are 13 points, the LMS estimator is based on the narrowest strip that contains 7 points. If the 13th data point is at Position A, then the LMS estimator will produce Line A as the estimated line. However, if there is a small error in the recording of the data and the 13th data point is recorded at Position B, then the LMS estimator will produce Line B as the estimated line. Thus, as was illustrated by Hettmansperger and Sheather (1992), the LMS estimator of the slope can be sensitive to small errors in the data. The instability of this estimator was discussed by Stromberg (1993), who produced software (Stromberg (1992)) that should detect the presence of instability in the estimates for some data sets. Ryan (1997) has also questioned the wisdom of using the LMS estimator. The instability of the LMS estimator is problematic, especially for an estimator that was designed to be robust.

Another robust regression estimator is the LTS estimator. This estimator minimizes the sum of the squared residuals, not including a portion of residuals with the largest squared values. If 25% trimming is used, the LTS line is the line that minimizes the squared residuals for the $\lfloor .75n \rfloor + 1$ observations with the smallest squared residuals, where $\lfloor x \rfloor$ is the greatest integer less than or equal to x. This is a reasonable procedure because we would like to trim off the outliers while retaining a large proportion of other observations in the regression fit. Note that the LTS estimator with 50% trimming minimizes the sum of squared residuals for the $\lfloor .5n \rfloor + 1$ observations with the smallest squared residuals, whereas the LMS estimator minimizes the median of the squared residuals.

Although the LTS estimator has a certain appeal, some problems can arise when finding the estimates. One problem is determining the amount of trimming that should be used. Should the trimming proportion be 25% or 10%? Since the amount of trimming may depend on the data set, the user may resort to an informal adaptive trimming procedure.

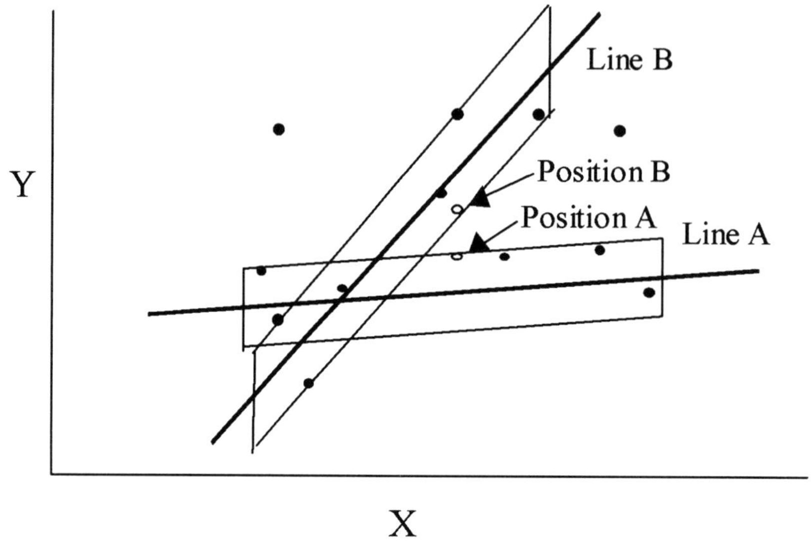

Figure 7.2. *A scatterplot that illustrates the sensitivity of the LMS estimator to small changes in the data.*

Another problem is the difficulty of computing the LTS estimates with moderate sized data sets. For example, if we trim 25% of $n = 50$ points, we minimize the sum of squared residuals for $\lfloor .75(50) \rfloor + 1 = 38$ observations with the smallest squared residuals, so we may need to evaluate $\binom{50}{38} = 1.2 \times 10^{11}$ data sets to determine the LTS estimator. Even if we can compute the LTS estimate, we may be hesitant to use it because there are few finite sample results comparing the LTS estimate to the OLS estimate. Consequently, there are a variety of problems with the LTS estimator that limit its usefulness.

Many of the robust regression approaches developed over the last 30 years are described in Rousseeuw and Leroy (1987). One of the more well-known robust regression methods is based on M-estimation. M-estimators minimize $\sum \rho\,(r_i)$, where r_i is the ordinary residual and ρ is a symmetric function about zero. Several functions have been proposed for use in M-estimation, and each of these functions has a tuning parameter that must be chosen. For details, see Hampel et al. (1986, Chap. 2) and Huber (1973). We should note that these methods are not, in their basic forms, adaptive in nature. A function selected to robustify an analysis by reducing the influence of large residuals may not be efficient if it is used on regression models that have normal errors. In addition, the scarcity of published results on the finite sample properties of these estimation procedures is a serious drawback to their use. If more finite sample comparisons of these estimates were available, researchers would have better guidance concerning the choice of a symmetric function and tuning parameters.

Many other types of robust regression methods have been proposed, including L-estimators, S-estimators, and R-estimators. The interested reader is referred to the books by Hampel et al. (1986), Rousseeuw and Leroy (1987), and Ryan (1997). Some software for the computation of these robust estimators is given in Marazzi (1993). The variety of estimators, the computational complexity of these estimators, and the scarcity of published finite sample results make it difficult for a researcher to select an appropriate method. As a consequence, although a great deal has been written on the subject of robust regression, the methods are not widely used. We will take a different approach to robust regression estimation in this chapter.

7.1.3 Objectives of Adaptive Estimation

We seek an adaptive estimator that avoids many of the problems associated with the robust regression estimators. Our objective in adaptive regression is to develop an estimator that

1. uses information from all the data points if the error distribution appears to be approximately normal,
2. effectively downweights outliers so that their influence is limited,
3. is robust in the sense that small changes in the data will not greatly change the estimates,
4. can be computed easily so that a SAS macro can be used to perform the necessary calculation.

Several approaches to adaptive estimation have been proposed. Portnoy and Koenker (1989) give relative efficiencies for several adaptive estimators of the slope parameter. Yuh and Hogg (1988) propose an adaptive estimator of the slope in a simple linear regression. We will use a WLS approach to adaptive regression with weights determined by the data. The weighting procedure is similar to that used for testing and estimation. Consequently, we do not need to develop an entirely new weighting scheme in order to develop an adaptive estimator.

7.2 An Adaptive Estimator of the Mean of a Symmetric Distribution

7.2.1 The Adaptive WLS Estimator

We shall see that Hogg's adaptive estimator, which we described in section 7.1, has excellent properties. However, it would be desirable to have an adaptive estimator that is often more precise than Hogg's adaptive estimator and uses the WLS approach. For the proposed adaptive estimator let y_i denote the ith observation and $\tilde{y}$ denote the median of the observations $\{y_1, \ldots, y_n\}$. In order to find the appropriate weights, we begin by estimating the c.d.f. of a symmetric distribution. We ensure that the empirical c.d.f. is symmetric by "doubling" the centered observations to obtain the set $Y_D = \{y_1 - \tilde{y}, \ldots, y_n - \tilde{y}, -(y_1 - \tilde{y}), \ldots, (y_n - \tilde{y})\}$. Next we compute a robust estimate of the standard deviation as $\hat{\sigma} = IQR_D/1.349$, where IQR_D is the interquartile range of the centered and doubled data Y_D.

The centered and standardized observations are defined as

$$z_i = \frac{y_i - \tilde{y}}{\hat{\sigma}}$$

for $i = 1, \ldots, n$. We then double these centered and standardized observations to obtain the set

$$Z_D = \{z_1, \ldots, z_n, -z_1, \ldots, -z_n\}$$

and denote the ith element in Z_D as $z_{D,i}$. For the adaptive estimator we use the same smoothing procedure that we used with paired data in section 5.3. At a point z we define the smoothed c.d.f. as

$$\hat{F}_h\left(z; Z_D\right) = \frac{1}{2n} \sum_{i=1}^{2n} \Phi\left(\frac{z - z_{D,i}}{h}\right),$$

where $\Phi\left(\cdot\right)$ is the c.d.f. of the standard normal distribution and $h = 1.26n^{-1/3}$ is the smoothing constant. In order to weight the observations, we need to compare z_i to the corresponding normal score, which is given by $\Phi^{-1}[\hat{F}_h(z_i;\ Z_D)]$. The weights are determined by

$$w_i = \frac{\Phi^{-1}\left[\hat{F}_h\left(z_i; Z_D\right)\right]}{z_i}$$

for $i = 1, \ldots, n$. If the differences are normally distributed, then $\Phi^{-1}[\hat{F}_h(z_i;\ Z_D)]$ should approximate z_i for all observations, which will produce weights near one. An outlier that produces an extremely large or small value of z_i will be given a small weight. Consequently, apart from the centering of the observations, the weighting procedure is quite similar to the one used in Chapter 5.

After the weights are computed they are applied to the observations to obtain the adaptive estimate. We noted in section 5.3 that the WLS estimate of β in the model $y_i = \beta + \varepsilon_i$ is $\hat{\beta} = \sum_{i=1}^{n} w_i^2 y_i / \sum_{i=1}^{n} w_i^2$. Consequently, $\hat{\beta}$ is our adaptive estimator of the mean of a symmetric distribution. If $w_i = 1$ for $i = 1, \ldots, n$, then $\hat{\beta} = \bar{y}$, which is the OLS estimator.

7.2.2 An Example

A data set published by Newcomb (1891) contained $n = 66$ measurements on the passage time of light. The interesting aspect of this data set is that we now know it is 33.02, but the researcher did not know it at the time the measurements were made. A dot plot of these observations, which is given in Figure 7.3, clearly shows one extreme outlier of -44 and another of -2. A value of 26.21 was obtained as the mean of these observations. This data set was among several analyzed by Stigler (1977), who recommended using the 10% trimmed mean.

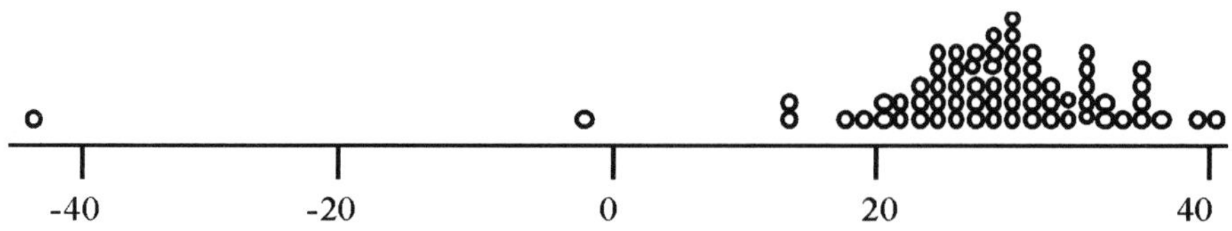

Figure 7.3. *Dot plot of* 66 *observations of the passage of light.*

In order to see how our adaptive estimator would work in this example, we used the SAS macro "estone" included in Appendix A. The SAS data set was called "light" and the variable was called "time." To obtain the adaptive estimator we used the following statement to call the SAS macro:

```
%estone(dataset=light, y=time);
```

The adaptive procedure gave a weight of 0.17 to the most extreme outlier −44 and a weight of 0.35 to the outlier −2. The other 64 observations had weights between 0.69 and 1.00. Thus, the adaptive estimator gave little weight to the most extreme observations. The adaptive estimate was 27.399, which is closer to the true value than the mean and nearly identical to the 10% trimmed mean of 27.4. Consequently, the adaptive estimator produced a more accurate estimate of the passage of time than the mean.

7.2.3 A Simulation Study

A simulation study was performed in order to compare the adaptive WLS estimator to Hogg's adaptive estimator, the 10% trimmed mean, and the mean. For this study we generated data from the uniform, beta, normal, logistic, 10% outlier, and 30% outlier distributions, which are the same distributions we used in Chapter 5. The densities for these six symmetric distributions were shown in Figures 5.2 to 5.7. We used sample sizes of $n = 10$, 20, and 50, and for each distribution and sample size we generated 10,000 data sets. These estimators are unbiased when used to estimate the mean of a symmetric distribution. The m.s.e.'s, which we will use as measures of the precision of these estimators, are tabulated in Table 7.1 for these four estimators.

We note, from the simulation results, that the m.s.e. of the mean is much greater than the m.s.e. of the other estimators when the observations were generated from the 10% outlier and 30% outlier distributions. The poor behavior of the mean has been noted by many statisticians and has motivated the search for better estimators. We also note that the m.s.e. of the mean is only slightly less than the m.s.e. of the other three estimators for normal errors.

The m.s.e. of the 10% trimmed mean is greater than the m.s.e. of the adaptive WLS estimator for observations from the 30% outlier distribution and from the uniform distribution. For $n \geq 20$ the m.s.e. of the 10% trimmed mean is equal to or greater than the m.s.e. of the adaptive WLS estimator for all distributions except the logistic distribution. Consequently, the adaptive WLS estimator appears to be preferred to the 10% trimmed mean.

Only the adaptive estimators have relatively low m.s.e. for all distributions. We note that for $n \geq 20$ the adaptive WLS estimator usually has a smaller m.s.e. than Hogg's estimator. While Hogg's adaptive estimator is remarkably effective considering the simplicity of the design, both adaptive estimators require software to compute the estimates. Some researchers may be more comfortable with Hogg's adaptive estimator because the component estimator selected by the selection statistic is easy to understand. Other researchers may be inclined to use the adaptive WLS estimator because it may be more precise when it is used with outlier distributions.

Table 7.1. *The estimated m.s.e. for four estimators of the mean of a symmetric distribution.*

	Distribution					
	Uniform	Beta	Normal	Logistic	10% Outliers	30% Outliers
			$n = 10$			
Adaptive WLS	0.131	0.119	0.107	0.097	0.155	0.527
Hogg's Adaptive	0.119	0.118	0.109	0.100	0.193	0.793
10% Trim. Mean	0.130	0.122	0.105	0.094	0.289	1.487
Mean	0.100	0.101	0.101	0.100	1.077	3.014
			$n = 20$			
Adaptive WLS	0.057	0.054	0.052	0.048	0.070	0.150
Hogg's Adaptive	0.055	0.055	0.053	0.051	0.089	0.185
10% Trim. Mean	0.069	0.061	0.052	0.046	0.099	0.603
Mean	0.051	0.050	0.049	0.051	0.548	1.538
			$n = 50$			
Adaptive WLS	0.015	0.019	0.021	0.019	0.027	0.048
Hogg's Adaptive	0.020	0.021	0.021	0.020	0.036	0.060
10% Trim. Mean	0.028	0.025	0.022	0.018	0.031	0.202
Mean	0.020	0.020	0.020	0.020	0.214	0.621

7.3 An Adaptive Estimator of Parameters in a Linear Model

In this section we will use a WLS approach to the estimation of parameters in a linear model. The weighting procedures are similar to those used in tests of significance and confidence intervals. Our objective is to develop an estimator that has a relatively small m.s.e. compared to the OLS estimator for a wide range of sample sizes and error distributions.

We will use the same multiple regression model used in previous chapters on testing and confidence intervals. That is, the multiple regression model is

$$y_i = \beta_0 x_{i,0} + \beta_1 x_{i,1} + \cdots + \beta_p x_{i,p} + \varepsilon_i$$

for $i = 1, \ldots, n$. Note that in previous chapters we made a distinction between complete and reduced models in order to perform a test of significance, but in this chapter the full model containing $p + 1$ independent variables will be used. We will write this model in matrix form as

$$Y = X\beta + \varepsilon,$$

where Y is the $n \times 1$ vector containing the dependent variable, X is an $n \times (p + 1)$ matrix containing the independent variables, β is the $(p + 1) \times 1$ vector of parameters to be estimated, and ε is the $n \times 1$ vector of errors. To include an intercept in the model we will use a column of ones as the first column of the X matrix.

We begin the adaptive estimation by computing the standardized deleted residuals

$$d_i = e_i \left[\frac{n-p-2}{SSE\,(1-h_{ii}) - e_i^2} \right]^{\frac{1}{2}}$$

for $i = 1, \ldots, n$. Note that these standardized deleted residuals are computed in the same manner as in Chapters 2 and 3, except that we use all $p+1$ independent variables in this model. We will weight the observations so that the c.d.f. of the studentized deleted residuals, after weighting, will approximate the c.d.f. of the t distribution with $\nu = n - (p+1) - 1 = n - p - 2$ degrees of freedom, which will be denoted by $T_{n-p-2}(\cdot)$. We then smooth the c.d.f. of these standardized deleted residuals by using a normal kernel with a bandwidth of $h = 1.587\,\hat{\sigma}\,n^{-1/3}$, as suggested by Polansky (1998). Since the observations are studentized the variance should not depart too much from $\sigma^2 = 1$, so a value of $h = 1.587\,n^{-1/3}$ is used to obtain the smoothed distribution function. This is the same smoothing procedure as we used for the adaptive test. Our objective is to weight the observations in such a way that the smoothed c.d.f. of the deleted residuals for the weighted model will agree with the c.d.f. of the t distribution. Let $D = \{d_1, \ldots, d_n\}$ be the set of studentized deleted residuals. The smoothed c.d.f. at point d over the set of all studentized deleted residuals (D) is computed as

$$\hat{F}_h(d; D) = \frac{1}{n} \sum_{i=1}^{n} \Phi\left(\frac{d - d_i}{h}\right),$$

where $\Phi(\cdot)$ is the c.d.f. of the standard normal distribution. After the smoothed c.d.f. of these studentized deleted residuals is obtained, we center the studentized deleted residuals by subtracting the estimated median $\tilde{d}$, which is determined by a search process so that $\hat{F}_h(\tilde{d}; D) = .5$. The centered studentized deleted residuals are calculated as $d_{c,i} = d_i - \tilde{d}$ for $i = 1, \ldots, n$ and will be called the residuals. The set of residuals will be denoted by $D_c = \{d_{c,1}, \ldots, d_{c,n}\}$, and we will let $t_i = T_{n-p-2}^{-1}(\hat{F}_h(d_{c,i}; D_c))$. To weight the observations we use

$$w_i = \frac{t_i}{d_{c,i}}$$

for $i = 1, \ldots, n$. If the error terms are normally distributed, then the smoothed c.d.f. of the centered studentized deleted residuals should approximate the c.d.f. of the t distribution with $\nu = n - p - 2$ degrees of freedom and the weights should approximate one. If the ith observation is an outlier, then $d_{c,i}$ will be large relative to t_i so that the ith observation will be given a small weight.

After we have computed the weights w_i for $i = 1, \ldots, n$, they can be used as the diagonal elements in the weighting matrix W with zero off-diagonal elements. We perform the WLS regression by premultiplying both sides of the model by W to obtain

$$WY = WX\beta + W\varepsilon.$$

This can be written as the transformed model

$$Y^* = X^*\beta + \varepsilon^*,$$

where $Y^* = WY$, $X^* = WX$, and $\varepsilon^* = W\varepsilon$. Using the transformed data, OLS methods are used to compute the parameter estimates. This weighting scheme is designed to reduce the influence of observations that have extreme studentized deleted residuals. If the error terms are normally distributed, the weights will approximate one and the estimates should approximate the OLS estimates.

7.4 An Adaptive Estimator of the Slope in a Simple Linear Regression

7.4.1 An Example

As a part of a study performed to investigate the relationship between food intake and body build, Jiang and Hunt (1983) measured the BMI (body weight in kilograms/height in square meters) and energy density (kcal/g) of foods over a 7-day period. These data are shown in Figure 7.4. Using BMI as the dependent variable and energy density as the independent variable in a simple linear regression we obtain the OLS fitted line $\hat{y} = 18.219 + 5.952x$. The OLS regression line shown as a solid line in Figure 7.4 gives a poor fit to the data. In contrast, the adaptive WLS regression line, which is $\hat{y} = 19.384 + 3.888x$, fits most of the data reasonably well, except for the outlier. The data were in a SAS data set named "fooddata," the dependent variable was named "bmi," and the independent variable was named "density." The adaptive estimator was obtained by using the "pointest" macro in Appendix A. The macro was invoked with

```
%pointest(dataset=fooddata, y=bmi, x=density);
```

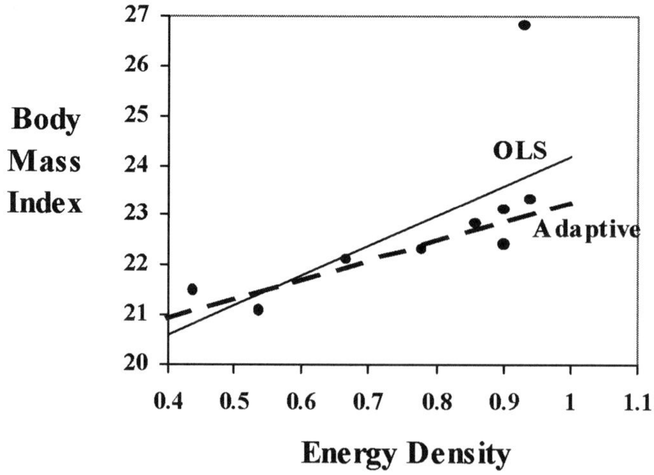

Figure 7.4. *A simple linear regression of BMI on energy density.*

The adaptive WLS procedure gave a weight of 0.182 to the outlier, while it gave weights between 0.920 and 1.158 for the other observations. Thus, the adaptive method automatically downweighted the outlier to produce the adaptively fitted line. Although the adaptive WLS regression line appears to fit these data, we will use a simulation study to evaluate the effectiveness of the adaptive estimation procedure.

7.4.2 A Simulation Study for the Adaptive Estimator

We began our evaluation of the adaptive estimator by using a simulation study to estimate the m.s.e. of the adaptive estimator of the slope in a simple linear regression. In these simulations we used data sets having $n = 10, 20$, and 50 observations with errors generated from one of the generalized lambda distributions (Ramberg et al. (1979)). These are the same distributions as have been used in many of the simulation studies in previous chapters. In these simulations the independent variable for the ith observation $(x_{i,1})$ was generated from a skewed distribution with $\alpha_3 = 1$ and $\alpha_4 = 5.4$, and the ith dependent variable was obtained by adding the error term from one of the generalized lambda distributions to $\beta_0 + \beta_1 x_{i,1}$. In each simulation 10,000 data sets were generated to estimate the m.s.e. of the adaptive estimator and the OLS estimator. The results of these simulations are tabulated in Table 7.2 along with the relative efficiency, which is defined as

$$\text{Relative Efficiency} = \frac{\text{estimated m.s.e. of the OLS estimator}}{\text{estimated m.s.e. of the adaptive estimator}}.$$

Relative efficiencies greater than one indicate that the adaptive estimator will, on average, provide a more accurate estimate of the slope parameter.

The relative efficiencies in Table 7.2 demonstrate that for $n \geq 20$ the adaptive estimator is more efficient than the OLS estimator for nonnormal error distributions. The simulations

Table 7.2. *The estimated m.s.e.'s for simple linear regression estimators of the slope. The independent variables were generated from a skewed distribution with $\alpha_3 = 1.0$ and $\alpha_4 = 5.4$.*

	Distributions								
Skewness	0	0	0	1	1	1	2	2	2
Kurtosis	1.8	3.0	6.0	4.2	5.4	8.4	11.4	12.6	15.6
					$n = 10$				
Adaptive WLS	0.176	0.180	0.157	0.151	0.156	0.149	0.128	0.132	0.133
OLS	0.169	0.169	0.167	0.163	0.166	0.166	0.168	0.167	0.167
Relative Eff.	0.96	0.94	1.06	1.08	1.07	1.11	1.31	1.27	1.26
					$n = 20$				
Adaptive WLS	0.060	0.067	0.061	0.054	0.056	0.054	0.042	0.045	0.045
OLS	0.064	0.065	0.066	0.064	0.065	0.064	0.065	0.065	0.065
Relative Eff.	1.07	0.96	1.10	1.19	1.16	1.19	1.55	1.44	1.42
					$n = 50$				
Adaptive WLS	0.016	0.023	0.019	0.017	0.018	0.018	0.012	0.013	0.014
OLS	0.021	0.022	0.022	0.022	0.022	0.022	0.022	0.022	0.022
Relative Eff.	1.34	0.98	1.13	1.33	1.21	1.26	1.75	1.64	0.154

also demonstrate that the adaptive estimator is less efficient than the OLS estimator for normal error distributions. We see that for $n = 10$ there is a small advantage to using the adaptive estimator if the errors are skewed and a small disadvantage if the errors are normal or uniform. Thus, for $n \geq 20$ the adaptive estimator appears to be superior to the OLS estimator of the slope. In contrast, the adaptive estimator of the slope proposed by Lo and Han (1997) does not appear to have high relative efficiency for $n = 50$.

7.4.3 Two-Step Adaptation

We have seen that the adaptive estimator for the slope in a simple linear regression model often has a smaller m.s.e. than that of the traditional OLS estimator of the slope. We will now investigate whether we can improve the estimator by further adaptation.

In section 7.3 we used the matrix W to represent the weighting matrix used to transform the linear model $Y = X\beta + \varepsilon$. By premultiplying by W we obtained

$$WY = WX\beta + W\varepsilon,$$

which was written as the transformed model

$$Y^* = X^*\beta + \varepsilon^*,$$

where $Y^* = WY$, $X^* = WX$, and $\varepsilon^* = W\varepsilon$. We now consider the possibility of using an adaptive method with the transformed model $Y^* = X^*\beta + \varepsilon^*$ to obtain an improved estimator of β.

One approach is to use the same adaptive weighting procedure on the transformed data as we used with the original linear model. To see why a second step might be helpful consider the data in the scatterplot in Figure 7.5. Both of the observations A and B are outliers and should be downweighted, but since both A and B are influential observations the deleted residuals may not be as large as they would have been had there been only one outlier. Hence, these two points may not be appropriately downweighted by the first step of the adaptive method, which suggests that a second step may be useful.

In order to determine if the two-step adaptive method is beneficial, we used the two-step estimator in the same simulation study as was used to generate the data for Table 7.2. We computed the m.s.e. of the two-step adaptive estimator and its relative efficiency to the OLS estimator for simulation studies that correspond to the sample sizes and error distributions in Table 7.2.

For $n = 20$ the second step of the adaptive method appeared to be somewhat beneficial for the data sets generated from the three highly skewed distributions with $\alpha_3 = 2$, but the second step was slightly harmful for data sets generated from the normal distribution. For example, with $\alpha_3 = 0$ and $\alpha_4 = 3.0$ we obtained a relative efficiency of 0.93 with the two-step estimator, which is slightly less than the value of 0.96 that we obtained with the one-step adaptive estimator. This suggests that it may be unwise to complicate the estimation of the weights when the errors are normally distributed. With $\alpha_3 = 1$ and $\alpha_4 = 5.4$ we obtained a relative efficiency of 1.16 for the two-step estimator, which is identical to the one-step adaptive result, and with $\alpha_3 = 2$ and $\alpha_4 = 12.6$ we obtained a relative efficiency of 1.51 for the two-step estimator, which is slightly larger than the 1.44 that we found with our

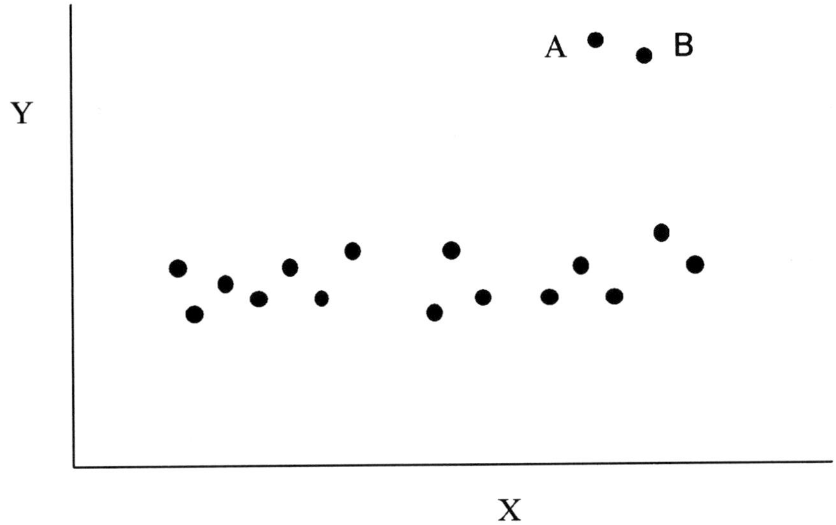

Figure 7.5. *Hypothetical data used to illustrate the difficulties associated with weighting two influential points.*

one-step adaptive estimator. The results for $n = 50$ were similar to those for $n = 20$. Thus, it appears that the second step is not always helpful in improving the estimates and can sometimes generate less precise estimates. Consequently, we cannot recommend the use of the two-step adaptive estimator.

7.5 An Adaptive Estimator of One Parameter in a Multiple Regression Model

7.5.1 An Example

The data shown in Table 7.3 are from a study by Coleman et al. (1966) that concerned the relationship of several factors to the mean verbal test scores of sixth graders (y). The data, which were also analyzed by Mosteller and Tukey (1977) and by Rousseeuw and Leroy (1987), include staff salaries per pupil (x_1), percent of white collar fathers (x_2), socioeconomic status composite deviation (x_3), mean of the teachers' verbal test scores (x_4), and mean of the mothers' educational level (x_5). These five independent variables were used to fit an OLS regression line of

$$\hat{y} = 19.95 + 1.79x_1 + 0.0436x_2 + 0.556x_3 + 1.11x_4 - 1.81x_5.$$

After making a SAS data set "coleman" we found the adaptive WLS estimates by invoking the macro

```
%pointest(dataset=coleman, y=y, x=x1 x2 x3 x4 x5);
```

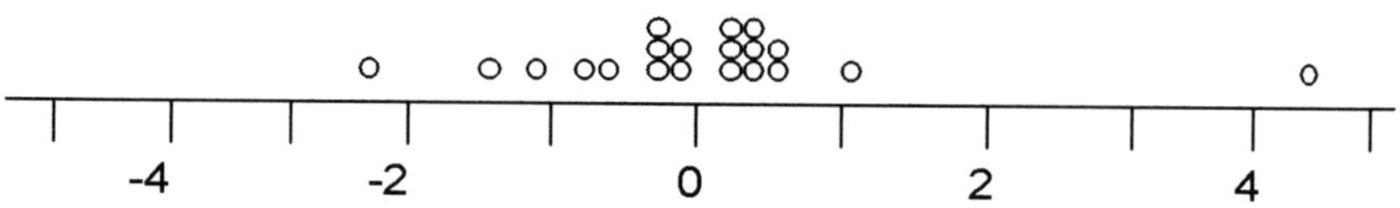

Figure 7.6. *Dot plot of studentized deleted residuals for the verbal scores of sixth graders.*

The adaptive WLS procedure produced the studentized deleted residuals shown in the dot plot in Figure 7.6. Because $d_{18} = 4.56$ the 18th observation was downweighted by the adaptive procedure. The weights for all observations, which are shown in the last column of Table 7.3, are near one except for observation 18, which was given a weight of $w_{18} = 0.473$. After weighting the observations the adaptive WLS method produced the fitted line of

$$\hat{y} = 29.07 - 1.67x_1 + 0.0730x_2 + 0.638x_3 + 1.12x_4 - 3.63x_5.$$

Although it is impossible to know if the adaptive WLS estimates are more accurate than the OLS estimates, we observe that they are in good agreement with a robust analysis of the same data given by Rousseeuw and Leroy (1987, Chap. 3). In their analysis they eliminated observations 3, 17, and 18 before performing OLS regression on the remaining

Table 7.3. *Data on the mean verbal scores of sixth graders in* 21 *schools.*

School	y	x_1	x_2	x_3	x_4	x_5	Weights
1	37.01	3.83	28.87	7.20	26.60	6.19	1.10320
2	26.51	2.89	20.10	−11.71	24.40	5.17	1.07375
3	36.51	2.86	69.05	12.32	25.70	7.04	0.89631
4	40.70	2.92	65.40	14.28	25.70	7.10	1.07255
5	37.10	3.06	29.59	6.31	25.40	6.15	1.10983
6	33.90	2.07	44.82	6.16	21.60	6.41	1.08662
7	41.80	2.52	77.37	12.70	24.90	6.86	1.10932
8	33.40	2.45	24.67	−0.17	25.01	5.78	1.07482
9	41.01	3.13	65.01	9.85	26.60	6.51	1.10804
10	37.20	2.44	9.99	−0.05	28.01	5.57	1.09983
11	23.30	2.09	12.20	−12.86	23.51	5.62	0.97483
12	35.20	2.52	22.55	0.92	23.60	5.34	1.08570
13	34.90	2.22	14.30	4.77	24.51	5.80	1.03912
14	33.10	2.67	31.79	−0.96	25.80	6.19	1.07822
15	22.70	2.71	11.60	−16.04	25.20	5.62	0.99705
16	39.70	3.14	68.47	10.62	25.01	6.94	1.11211
17	31.80	3.54	42.64	2.66	25.01	6.33	1.02109
18	31.70	2.52	16.70	−10.99	24.80	6.01	0.47254
19	43.10	2.68	86.27	15.03	25.51	7.51	1.11248
20	41.01	2.37	76.73	12.77	24.51	6.96	1.09876

observations. Using this approach they obtained a fitted line of

$$\hat{y} = 29.75 - 1.203x_1 + 0.082x_2 + 0.659x_3 + 1.098x_4 - 3.898x_5.$$

Except for $\hat{\beta}_1$, which they judged to be the least significant independent variable, the robust estimates are in close agreement with the adaptive WLS estimates. Of course, it is unwise to make a firm conclusion based on the analysis of this data set because we do not know the true values of the regression coefficient. In order to fully evaluate the effectiveness of the adaptive WLS procedure with multiple regression we will use a simulation study.

7.5.2 A Simulation Study for the Adaptive Multiple Regression Estimator

In order to determine the effectiveness of the adaptive estimator we used a simulation study with a multiple regression model with error distributions and sample sizes identical to those used in the simulation for the simple linear regression. In these simulations we used the model with two independent variables

$$y_i = \beta_0 + \beta_1 x_{i,1} + \beta_2 x_{i,2} + \varepsilon_i$$

for $i = 1, \ldots, n$. The two independent variables were generated from a bivariate lognormal distribution with a mean of one for both variables, a standard deviation of one for both variables, and a correlation of $\rho = 0.8$. This bivariate distribution was used because we wanted to see if the adaptive estimator would be effective if the independent variables had skewed distributions with a moderate amount of multicollinearity. After x_1 and x_2 were generated, the y-value was obtained by adding an error to $\beta_0 + \beta_1 x_{i,1} + \beta_2 x_{i,2}$, where the errors were generated from one of the generalized lambda distributions. The estimated m.s.e.'s and the relative efficiencies are given in Table 7.4 for the adaptive and OLS estimators of β_2.

These simulation results show that, if 20 or more observations are used for estimating β_2, the adaptive estimator has a lower m.s.e. than the OLS estimator for nonnormal error distributions. For $n = 10$ observations the adaptive estimator has a smaller m.s.e. for skewed distributions, but the relative efficiencies are close to one. Although the m.s.e.'s for the multiple regression estimator are much larger in Table 7.4 than they are for the simple linear regression estimator in Table 7.2, due primarily to the high correlation between the independent variables in these simulations, the relative efficiencies in Table 7.4 are quite similar to those in Table 7.2. It appears that for $n \geq 20$ the adaptive estimator is superior to the OLS estimator if the distributions are nonnormal.

7.6 Concluding Remarks on Adaptive Estimation

We have seen that the adaptive estimators of the mean of a symmetric distribution often are more precise than the mean. We also noted that the 10% trimmed mean, which often has a smaller m.s.e. than the mean, has a greater m.s.e. than the adaptive estimators for certain long-tailed distributions. Consequently, only the adaptive estimators can be recommended for estimating the mean of a symmetric distribution. Of these, Hogg's estimator

Table 7.4. *The estimated m.s.e.'s for multiple linear regression estimators of* β_2. *The independent variables were generated from a bivariate lognormal distribution with a correlation of* 0.8.

	Distributions								
Skewness	0	0	0	1	1	1	2	2	2
Kurtosis	1.8	3.0	6.0	4.2	5.4	8.4	11.4	12.6	15.6
					$n = 10$				
Adaptive WLS	1.37	1.49	1.30	1.33	1.32	1.27	1.12	1.15	1.09
OLS	1.30	1.40	1.35	1.37	1.39	1.38	1.35	1.38	1.35
Relative Eff.	0.95	0.94	1.04	1.03	1.05	1.08	1.21	1.20	1.24
					$n = 20$				
Adaptive WLS	0.334	0.367	0.326	0.313	0.321	0.304	0.256	0.250	0.262
OLS	0.346	0.356	0.353	0.360	0.363	0.359	0.361	0.348	0.353
Relative Eff.	1.04	0.97	1.08	1.15	1.13	1.18	1.41	1.39	1.35
					$n = 50$				
Adaptive WLS	0.072	0.092	0.081	0.072	0.073	0.078	0.056	0.058	0.062
OLS	0.088	0.090	0.090	0.091	0.086	0.092	0.089	0.088	0.090
Relative Eff.	1.22	0.98	1.11	1.27	1.18	1.18	1.60	1.52	1.44

has excellent performance and the statistics that are components of this estimator are easy to understand. However, the adaptive WLS estimator often has a slightly smaller m.s.e. than Hogg's estimator, and it uses the same general weighting scheme used throughout this book for tests of significance and confidence intervals. The choice is one of personal preference since they both seem to have excellent performance compared to the other estimators.

The adaptive WLS estimator for the parameters in a multiple regression model has been shown to be superior to the OLS method if the distribution of errors is nonnormal. In addition, it is surprising to see how little is lost in terms of efficiency when the adaptive estimator is used with normal errors. In general, there appears to be an advantage to using the adaptive estimators whenever $n \geq 20$ if the error distributions are not known to be normal.

Chapter 8

Additional Adaptive Methods and Special Topics

8.1 Rank-Based Tests for the One-Way Layout

Rank-based adaptive tests played an important role in the development of adaptive methods because it could easily be demonstrated that these tests maintained their significance level. Over the last 10 years several rank-based adaptive tests for the one-way layout have been proposed. The method proposed by Büning (1996), which will be described in section 8.1.2, is an extension and modification of the two-sample adaptive test proposed by Hogg, Fisher, and Randles (1975). Another rank-based test for the one-way layout was proposed by O'Gorman (1997a), who used a continuous method of adapting the rank scores. These tests maintain their significance level because they use selection statistics based on the order statistics of the combined data and test statistics that are functions of the ranks of the combined data. Because they maintain their significance level, these adaptive tests address the concerns of those researchers who feel that adaptive procedures may not be valid. In addition, the rank-based adaptive tests show that adaptive procedures are potentially valuable because they can increase the power of certain tests. Before we describe these tests we will show that rank-based adaptive tests, if constructed properly, maintain their significance level.

8.1.1 Why Rank-Based Adaptive Tests Maintain Their Significance Level

In this section we will assume that we have obtained independent random samples from K groups. Let n_i be the number of observations in the ith group and let Y_{ij} be the jth observation in the ith group. Using this notation, the vector of observations can be written as $\left(Y_{11}, \ldots, Y_{1n_1}, \ldots, Y_{k1}, \ldots, Y_{kn_K}\right)$. We will assume that the shift model is appropriate so that the c.d.f.'s can be expressed as $F(x-\theta_1), \ldots, F(x-\theta_K)$. We will test the null hypothesis $H_0 : \theta_1 = \cdots = \theta_K$ versus the alternative $H_1 : \theta_i \neq \theta_j$ for at least one pair (i, j) with $i \neq j$.

In order to show that the adaptive test maintains its size, we begin by defining the vector of ranks $R = \left(R_{11}, \ldots, R_{1n_1}, \ldots, R_{k1}, \ldots, R_{Kn_K}\right)$, where R_{ij} is the rank of Y_{ij} in the combined sample of $n = n_1 + \cdots + n_K$ observations. Next, we find the conditional density of $Y_{(1)}, \ldots, Y_{(n)}$ for any specific vector of ranks, say, r. The conditional density is

$$g(y_{(1)}, \ldots, y_{(n)} \,|R = r) = \frac{\Pi_{i=1}^n f(y_{(i)})}{P[R = r]} = n! \prod_{i=1}^{n} f(y_{(i)})$$

for $y_{(1)} < \cdots < y_{(n)}$. We observe that this conditional density equals the marginal density of the order statistics $g(y_{(1)}, \ldots, y_{(n)})$. Since the conditional density equals the marginal density for all rank vectors, the order statistics are independent of the vector of ranks. See Randles and Wolfe (1979, Chap. 8) for details.

In an adaptive test we use a selection statistic Q, which may be a vector, to select a component test. If Q falls in the ith region, we use the test statistic S_i along with the critical value c_i to determine if we should reject the null hypothesis. We will now show that the adaptive test maintains its size for all distributions of errors if Q is a function of the order statistics and if the test statistics are based on ranks.

We assume that we have a finite number I of regions that will be used in the test and let E_i be the event that Q falls in the ith region. For any distribution of errors the overall probability of rejection of the null hypothesis is

$$P\,(\text{Rejection}) = \sum_{i=1}^{I} P\,(\text{Rejection} \cap E_i) = \sum_{i=1}^{I} P\,(S_i \geq c_i \cap E_i).$$

But if the test statistics are based on the vector R and if Q is a function of the order statistics, then, under the null hypothesis, the test statistics are independent of Q and

$$P\,(\text{Rejection}) = \sum_{i=1}^{I} P\,(S_i \geq c_i)\, P\,(E_i).$$

In each region we can always determine a c_i such that $P\,(S_i \geq c_i) \leq \alpha$. Thus, under the null hypothesis,

$$P\,(\text{Rejection}) \leq \sum_{i=1}^{I} \alpha P\,(E_i) = \alpha.$$

This result holds for all error distributions. Consequently, we can be assured that a rank-based adaptive test will maintain its significance level if the component tests use test statistics that are functions of the ranks of the combined sample, and the selection statistic is a function of the order statistics of the combined sample.

8.1.2 Büning's Adaptive Test

We now consider an adaptive test for the one-way layout proposed by Büning (1996). Let $\bar{U}_{.05}$ be the average of the $.05N$ largest observations in the combined sample, let $\bar{L}_{.05}$ be the average of the $.05N$ smallest observations, and let $\bar{M}_{.50}$ be the average of the $.5N$ middle observations. In Büning's adaptive test the choice of test statistic is based on a vector of selection statistics. The first selection statistic, which measures the asymmetry in the sample distribution, is

$$Q_3 = \frac{\bar{U}_{.05} - \bar{M}_{.5}}{\bar{M}_{.5} - \bar{L}_{.05}}.$$

If the empirical distribution is nearly symmetric, then Q_3 should approximate 1.0. If the empirical distribution is skewed to the right, then Q_3 may greatly exceed 1.0.

The second selection statistic measures the distance between the upper and lower tails of the sample distribution. It is defined as

$$Q_4 = \frac{\bar{U}_{.05} - \bar{L}_{.05}}{\bar{U}_{.5} - \bar{L}_{.5}},$$

where $\bar{U}_{.5}$ is the average of the largest $.5N$ observations and $\bar{L}_{.5}$ is the average of the smallest $.5N$ observations. For the uniform distribution $Q_4 = 1.9$, but for long-tailed distributions Q_4 may exceed 3. In the Büning test the vector $Q = (Q_3, Q_4)$ is used to select the most appropriate rank test statistic.

The adaptive test uses linear rank tests for the one-way layout as components of the overall test. In order to calculate the test statistic, we first combine the two samples and find the ranks of the observations over both samples. Often, in nonparametric statistics, we use the ranks themselves in the test statistic, as we do with the Wilcoxon rank-sum test. However, in some nonparametric tests we use some function of the ranks instead of the ranks themselves. The function of the rank is called the rank score, which will be denoted by $a(R_{ij})$ for an observation whose rank is R_{ij}. The component tests will be based on these rank scores.

If the distribution is not too highly skewed ($Q_3 < 2$) and has moderate tails ($2 \leq Q_4 \leq 3$), then the Kruskal–Wallis tests will be used. This region is illustrated in Figure 8.1. In the Kruskal–Wallis test, the scores are simply the ranks, so that

$$a_{KW}(i) = i \quad \text{if } 1 \leq i \leq n.$$

Other tests can be constructed for special circumstances. If the observations appear to be from a light-tailed distribution, we could give relatively larger weights to the largest and smallest ranks by using the following scores in a linear rank test statistic:

$$a_{LT}(i) = \begin{cases} i - (n+1)/4 & \text{if } i \leq (n+1)/4, \\ 0 & \text{if } (n+1)/4 < i < 3(n+1)/4, \\ i - 3(n+1)/4 & \text{if } i \geq 3(n+1)/4. \end{cases}$$

These scores were among those suggested by Gastwirth (1965).

If the observations are from a heavy-tailed distribution, Büning suggested using the following set of scores:

$$a_{HT}(i) = \begin{cases} -n/4 + 1 & \text{if } i < (n/4) + 1, \\ i - (n+1)/2 & \text{if } (n/4) + 1 \leq i \leq 3(n+1)/4, \\ n/4 + 1 & \text{if } i > 3(n+1)/4. \end{cases}$$

If the sample distribution appears to be skewed to the right, which is a common occurrence with many real-world data sets, we could use a test that is sensitive to right-skewed data.

One such set of scores is

$$a_{RS}(i) = \begin{cases} i - (n+1)/2 & \text{if } i \leq (n+1)/2, \\ 0 & \text{otherwise.} \end{cases}$$

For the ith region the test statistic is

$$S_i = \frac{\sum_{i=1}^{K} n_i \left(\bar{a}_i - \bar{a}\right)^2}{\left[\sum_{i=1}^{K} \sum_{j=1}^{n_i} \left\{a\left(R_{ij}\right) - \bar{a}\right\}^2\right] \Big/ (n-1)},$$

where

$$\bar{a}_i = \left[\sum_{j=1}^{n_i} a(R_{ij})\right] \Big/ n_i$$

and

$$\bar{a} = \left[\sum_{i=1}^{K} \sum_{j=1}^{n_i} a\left(R_{ij}\right)\right] \Big/ n.$$

If ties are present, the average score method is used to assign scores. Conover (1973, Thms. 4.2 and 4.5) showed that this statistic has asymptotically a chi-square distribution with $K - 1$ degrees of freedom, under the null hypothesis, if average scores are used for ties, provided the score function satisfies certain conditions. These conditions are satisfied by the score functions used in Büning's test.

In the Büning test, the statistics Q_3 and Q_4 are used to determine the appropriate set of rank scores. If $Q_3 \leq 2$ and $2 \leq Q_4 \leq 3$, the Kruskal–Wallis scores $\{a_{KW}(i),\ i = 1, \ldots, n\}$ will be used to compute the test statistic. If the distribution is not too skewed and has light tails such that $Q_3 \leq 2$ and $1 \leq Q_4 < 2$, then the scores for light-tailed distributions $\{a_{LT}(i),\ i = 1, \ldots, n\}$ will be used to compute the test statistic. If $Q_4 > 3$, the scores for the heavy-tailed distributions $\{a_{HT}(i),\ i = 1, \ldots, n\}$ will be used to compute the test statistic. If the distribution appears to be so skewed that $Q_3 > 2$ and if $Q_4 \leq 3$, then the scores for right-skewed distributions $\{a_{RS}(i),\ i = 1, \ldots, n\}$ will be used to compute the test statistic. These regions and their corresponding scores are indicated in Figure 8.1.

In order to perform the Büning test, we compute the selection statistics Q_3 and Q_4 that are used to select a set of rank scores. These scores are then used to compute the test statistic. We then use the fact that the test statistic has approximately a χ^2 distribution with $\nu = K - 1$ degrees of freedom to decide if we should reject the null hypothesis.

Büning used a Monte Carlo simulation to estimate the power of this adaptive test. He used $K = 3$ groups with $n_1 = n_2 = n_3 = 10$ and generated 1000 data sets for each of six error distributions. The distributions included the uniform, the normal, the double-exponential, the Cauchy, a contaminated normal, and the exponential. The simulations showed, as expected, that the test maintained its significance level. The simulation study demonstrated that the adaptive test had greater power than the F test for nonnormal distributions. For example, a simulation using exponential errors produced empirical powers of 52.7% for the F test, 68.5% for the Kruskal–Wallis test, and 73.7% for the adaptive test. When normal errors were used the adaptive test was nearly as powerful as the F test. One simulation using normal errors produced empirical powers of 42.2% for the F test, 38.0%

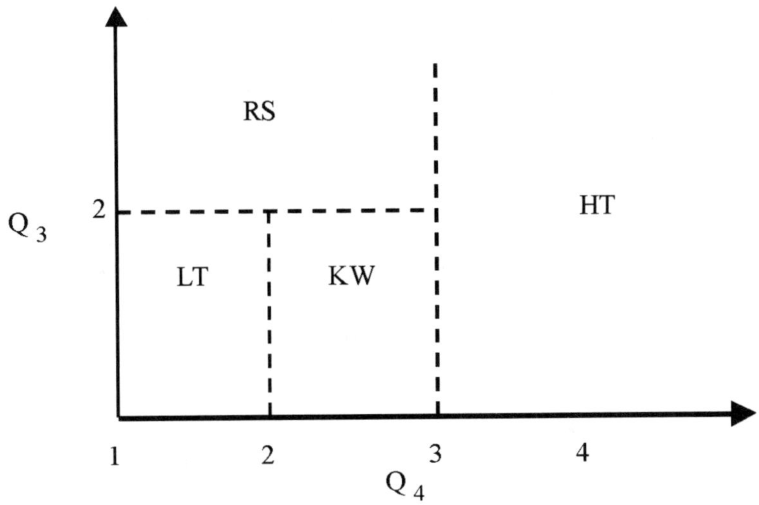

Figure 8.1. *Regions determined by Q_3 and Q_4 that are used to select the rank scores for the Büning test.*

for the adaptive test, and 39.3% for the Kruskal–Wallis test. Consequently, although only one sample size was used in the published simulations, it seems reasonable to conclude that Büning's test will be an effective and fair adaptive test if the sample sizes equal or exceed 10 in each group.

8.1.3 A Continuous Rank-Based Test

One disadvantage of the Büning test is that, if the selection vector (Q_3, Q_4) is near a boundary that determines which scores are to be used, a small change in the data may move (Q_3, Q_4) over a boundary of one of the regions shown in Figure 8.1. Thus, a small change in the data may produce a large change in the test statistic. To avoid this problem for the two-sample test, Ruberg (1986) proposed a continuously adaptive test based on the same selection statistics used in Hogg's adaptive test. In Ruberg's test the rank scores change only slightly when small changes are made to the data. Unfortunately, the Ruberg test is rather complex, and O'Gorman (1996) has shown that it can have relatively low power in certain situations.

O'Gorman (1997a) proposed a continuously adaptive test using selection statistics based on sample percentiles. In that test the sample percentiles were used to describe the lengths of the right and left tails of the empirical distribution. For convenience, let $L = (n+1)/4$ and $U = 3(n+1)/4$. The basic idea was to make the scores for those observations falling above the 75th percentile, where $R_{ij} > U$, depend on the length of the right tail of the distribution and the scores for the observations falling below the 25th percentile, where $R_{ij} < L$, depend on the length of the left tail of the distribution. The scores for the continuously adaptive test statistic use the Wilcoxon scores for the observations falling between the 25th and 75th percentiles.

The tail lengths are defined using the sample percentiles. Let $\hat{\xi}_p$ be the sample $100p$th percentile so that the interquartile range can be written as $IQR = \hat{\xi}_{.75} - \hat{\xi}_{.25}$. The right tail length measure is defined as $T_R = (\hat{\xi}_{.95} - \hat{\xi}_{.75})/IQR$ and the left tail length measure as $T_L = (\hat{\xi}_{.25} - \hat{\xi}_{.05})/IQR$. For two-sample tests we want our rank scores to approximate the Wilcoxon scores, as shown in Figure 8.2, if the distributions are logistic since it can be shown (Randles and Wolfe (1979, p. 300)) that the Wilcoxon test is the locally most powerful rank test for detecting a shift with the logistic distribution. For long-tailed distributions we want the rank scores to de-emphasize the extreme ranks in the manner shown in Figure 8.3. For short-tailed distributions we want the rank scores to emphasize the extreme ranks in the manner shown in Figure 8.4.

Before defining this continuously adaptive test, we note that the logistic distribution, which has an inverse c.d.f. of $F^{-1}(p) = \ln\left[p/(1-p)\right]$, has a right tail length measure of

$$\frac{F^{-1}(.95) - F^{-1}(.75)}{F^{-1}(.75) - F^{-1}(.25)} = \frac{\ln(19) - \ln(3)}{\ln(3) - \ln(1/3)} = 0.8401.$$

We also note that, if the distribution is logistic, the right tail length measure equals the left tail length measure. Because we want our scores to approximate the Wilcoxon scores

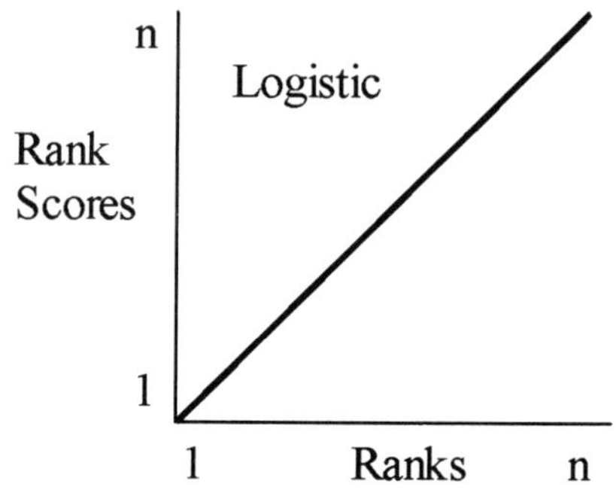

Figure 8.2. *Scores for the proposed test for a logistic distribution.*

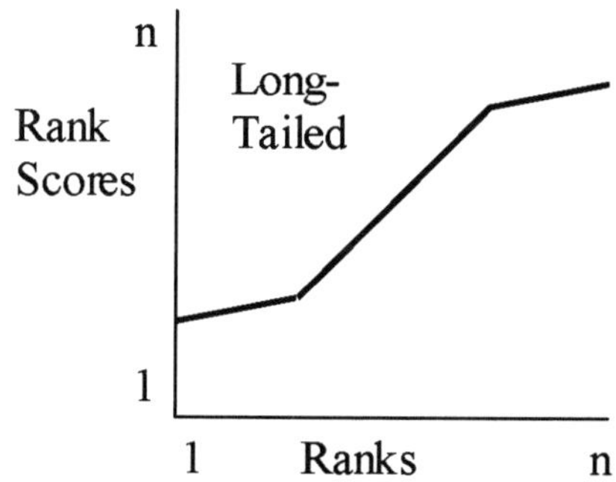

Figure 8.3. *Scores for the proposed test for a long-tailed distribution.*

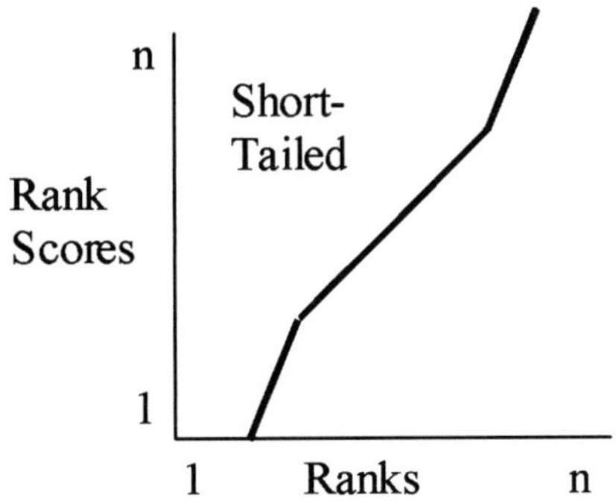

Figure 8.4. *Scores for the proposed test for a short-tailed distribution.*

if the tail lengths approximate those of the logistic distribution, we divide the tail length measures of the logistic distribution by the tail length measures of the empirical distribution to calculate the slopes of the score function.

One difficulty with the continuously adaptive approach is that if n is small, very large scores can be assigned to the observations in the tails if the largest (or smallest) observations are grouped together in a way that makes T_R (or T_L) very small. Consequently, in order to improve the small sample performance of this test, a minimum tail length of 0.4 is used. This minimum tail length corresponds to that obtained from a uniform distribution, which has tail lengths of 0.4. Thus, the tail length measures used in the continuously adaptive test are $T_L^* = \max(T_L, 0.4)$ and $T_R^* = \max(T_R, 0.4)$. Since the logistic distribution has a tail length measure equal to 0.8401 and since we want to approximate the Wilcoxon scores if our sample has the same tail length as the logistic distribution, we define the rank scores for the continuously adaptive test as

$$a_{CA}(R_{ij}) = \begin{cases} L + \left(0.8401/T_L^*\right)^2 (R_{ij} - L) & \text{if } R_{ij} < L, \\ R_{ij} & \text{if } L \le R_{ij} \le U, \\ U + \left(0.8401/T_R^*\right)^2 (R_{ij} - U) & \text{if } R_{ij} > U. \end{cases}$$

If we took a large sample from a logistic distribution, then T_L^* and T_R^* would approximate 0.8401. Hence, the rank scores would lie on a line with a slope of $\left(\frac{0.8401}{0.8401}\right)^2 = 1$ in both tails. Thus, the rank scores would approximate the Wilcoxon scores. A long-tailed distribution with $T_L = T_R = 2$ would have rank scores on a line with slopes of $\left(\frac{0.8401}{2}\right)^2 = 0.176$ in both tails. This will have the effect of downweighting the extreme ranks. The ratio of the tail lengths is squared because, after experimentation with some other values for the exponent, O'Gorman (1996) found that squaring the ratio produced high power for a variety of error distributions. He showed that this test has power generally equal to or greater than the nonadaptive two-sample tests.

For the one-way layout, a simulation study that was used to determine the effectiveness of this adaptive method was published by O'Gorman (1997a). The simulation showed that the test maintained its significance level and was more powerful than the F test with skewed distributions having $\alpha_3 = 1.0$. However, there was considerable loss in power with normal errors. For example, with $K = 3$ and $n_1 = n_2 = n_3 = 8$ the power of the adaptive test was 45% for a certain shift alternative, which is considerably less than the 52% power that was obtainable for the F test. Consequently, although this test maintains its size and is reasonably powerful when the errors are skewed, it performs poorly when the errors are approximately normally distributed. Since the adaptive test described in section 4.2 is nearly as powerful as this test with error distributions that are skewed, and because the adaptive WLS test is nearly as powerful as the F test with normally distributed errors, the adaptive WLS test is recommended over this rank-based adaptive test.

8.1.4 Adaptive Tests for Ordered and Umbrella Alternatives

In sections 8.1.2 and 8.1.3 we described the rank-based adaptive tests that could be used to test the null hypothesis $H_0 : \theta_1 = \cdots = \theta_K$ versus the alternative $H_a : \theta_i \neq \theta_j$ for at least one pair (i, j) with $i \neq j$. If the researcher has some prior information about the ordering of the alternatives, then a test designed for ordered alternatives may be more powerful than the traditional test. For example, if animals are assigned to K groups corresponding to K dosages of an experimental drug, an ordered alternative might be considered. If an ordered alternative is appropriate, we could use the Jonckheere–Terpstra statistic to test $H_0 : \theta_1 = \cdots = \theta_K$ versus the ordered alternative $H_a : \theta_1 \leq \cdots \leq \theta_K$ with at least one strict inequality. This nonadaptive nonparametric test is described by Randles and Wolfe (1979, Chap. 12).

An adaptive generalization of this test was proposed by Büning (1999). He noted that Jonckheere-type tests can be written using rank scores and proposed three sets of rank scores that could be used, in addition to the Wilcoxon scores. In his test for an ordered alternative, Büning uses the same selection statistics, Q_3 and Q_4, defined in section 8.1.2. These selection statistics are used to select one of the four rank scores that are then used in a Jonckheere-type test statistic. This test maintains its size and has good power compared to other tests for ordered alternatives. For details of this adaptive test and for comparisons with other tests designed for ordered alternatives, see Büning (1999).

Nonparametric tests have also been developed for umbrella alternatives. With these methods we test $H_0 : \theta_1 = \cdots = \theta_K$ versus the umbrella alternative $H_a : \theta_1 \leq \cdots \leq \theta_{l-1} \leq \theta_l \geq \theta_{l+1} \geq \cdots \geq \theta_K$ with at least one strict inequality, where l is some integer between 1 and K. A nonadaptive rank test for umbrella alternatives was given by Mack and Wolfe (1981). Büning and Kössler (1998) proposed an adaptive test, using the selection statistics Q_3 and Q_4 along with the rank scores defined in section 8.1.2, for the umbrella alternative. They showed that when l is known, the adaptive test maintains its size and is reasonably powerful compared to the Mack–Wolfe test.

8.2 Dealing with Unequal Variances

8.2.1 Departures from the Shift Model in Two-Sample Tests

In all of the simulations that we have performed in the previous chapters, we have assumed that a shift model was appropriate. For the two-sample test we generated data in the first sample from a distribution with a c.d.f. of $F(y)$ and for the second sample from a distribution with a c.d.f. of $F(y - \Delta)$, where Δ is a constant. We will call the observations generated from the first distribution the control observations and those from the second distribution the treatment observations. Note that if $\Delta > 0$, the c.d.f. of the treatment observations will be shifted to the right so that, in most of the data sets, the treatment observations will tend to be greater than the control observations. If the shift model is appropriate, the treatment observations should have approximately the same variability as the control observations. However, with real data, it is not unusual to find greater variability in the treatment data than in the control data because the treatment effects may be variable.

It should be pointed out that the failure to conform to a shift model is a potential problem for all tests, including the adaptive tests that we have described in this book. In order to evaluate the performance of tests of significance for variable treatment effects, we will use simulation studies that use a variable, rather than a constant, treatment effect.

When the shift model was used, a constant Δ was added to each variate generated to obtain an observation for the treatment group. As we have done in previous chapters, we will generate a random variate from a distribution with a c.d.f. of $F(y)$ to obtain an observation for the control group. In order to investigate the effectiveness of these tests when the treatment effect is variable, we will add a uniform variate generated from $U[0, 2\Delta]$ to a variate generated from a distribution with a c.d.f. of $F(y)$ in order to obtain an observation for the treatment group. By generating random treatment effects from a $U[0, 2\Delta]$ distribution we are using, on average, a treatment effect of Δ. This random treatment effect will produce greater variability in the treatment group than in the control group.

The power estimates in Table 8.1 are based on a simulation study using 10,000 data sets generated with variable treatment effects. The treatment effect parameter (Δ) was the same as was used in the constant treatment effects models described in section 2.7. As expected, the power of these tests with variable treatment effects is slightly less than the power observed with constant treatment effects. With $n_1 = n_2 = 20$ the adaptive test was somewhat more powerful than the t test when the data were generated from skewed distributions, but the power advantage of the adaptive test over the t test was not as great as we observed in Chapter 2 with constant treatment effects. The power estimates in Table 8.1 show that with variable treatment effects the adaptive test is more powerful than the t test for nonnormal errors and nearly as powerful as the t test for normal errors.

Table 8.1. *The empirical power in percent for the adaptive test and the pooled t test when variable treatment effects were used with a significance level of $\alpha = .05$.*

	Distributions								
Skewness	0	0	0	1	1	1	2	2	2
Kurtosis	1.8	3.0	6.0	4.2	5.4	8.4	11.4	12.6	15.6
				$n_1 = 10,\ n_2 = 10$					
Adaptive	43.3	43.4	48.1	47.8	47.5	49.7	57.2	56.2	53.9
t test	43.3	44.6	48.0	45.8	46.3	48.8	51.0	51.5	50.5
				$n_1 = 20,\ n_2 = 20$					
Adaptive	50.5	46.0	51.7	54.3	52.4	53.9	63.4	61.9	61.7
t test	46.6	47.0	49.1	47.9	48.5	50.0	50.5	50.7	51.9
				$n_1 = 48,\ n_2 = 12$					
Adaptive	49.9	44.5	50.6	54.6	52.2	53.4	64.4	63.0	61.5
t test	44.7	45.3	47.1	47.8	47.4	48.2	50.1	50.5	51.3

Based on these results we can recommend the adaptive test over the t test if the treatment effects have roughly the same sort of variability that we have used in these simulation studies. However, if the treatment effect is highly variable, the results shown in Table 8.1 may not reflect the performance of these tests. We now consider an appropriate test for highly variable treatment effects.

8.2.2 A Test for Location and Scale

We have seen that the adaptive test is reasonably powerful compared to the t test when the treatment effects are somewhat variable. By adding a variable treatment effect that was a $U[0, 2\Delta]$ variate we ensured that the distributions would be stochastically ordered. That is, if the data in the first sample were generated from $F(y)$ and the data in the second sample were generated from $G(y)$, we designed the simulation so that $F(y) \geq G(y)$ for all values of y.

However, there may be some situations where the treatment effect could be either positive or negative, so that the distributions may not be stochastically ordered. For example, a drug may cause some people to sleep more than usual and other people to sleep less than usual. In this situation the treatment observations may have much more variability than the control observations and the distributions may not be stochastically ordered. In this section we will describe tests that are sensitive to differences in location and variability.

If the researcher believes that there may be a change in location and scale, an excellent nonparametric two-sample test that is sensitive to differences in both location and variability was proposed by Lepage (1971). The test is based on a combination of a Wilcoxon test statistic used to detect location changes and an Ansari–Bradley statistic used to detect differences in variability.

Suppose we have n_1 observations in the first sample and n_2 observations in the second sample and, after combining these samples, we find the ranks of the observations in the first sample $\{R_i,\ i = 1, \ldots, n_1\}$. The Wilcoxon test statistic is based on the sum of the ranks in the first sample $W = \sum_{i=1}^{n_1} R_i$, and the Ansari–Bradley test is based on

$$A = \sum_{i=1}^{n_1} \left\{ \frac{n+1}{2} - \left| R_i - \frac{n+1}{2} \right| \right\}.$$

The Lepage test statistic is

$$L = \left(\frac{W - \mu_W}{\sigma_W} \right)^2 + \left(\frac{A - \mu_A}{\sigma_A} \right)^2,$$

where $\mu_W = n_1(n+1)/2$, $\sigma_W^2 = n_1 n_2 (n+1)/12$,

$$\mu_A = \begin{cases} n_1(n+2)/4 & \text{if } n \text{ is even,} \\ n_1(n+1)^2/(4n) & \text{if } n \text{ is odd,} \end{cases}$$

and

$$\sigma_A^2 = \begin{cases} n_1 n_2 (n^2 - 4)/[48\,(n-1)] & \text{if } n \text{ is even,} \\ n_1 n_2 (n+1)\left(n^2+3\right)/\left(48n^2\right) & \text{if } n \text{ is odd.} \end{cases}$$

The null hypothesis is that the first sample is from a distribution with a c.d.f. of $F\left[(y-\xi)/\sigma\right]$ and the second sample is from a distribution with a c.d.f. of $F\left[(y-\eta)/\tau\right]$. We will test $H_0: \xi=\eta,\ \sigma=\tau$ against the alternative $H_a: \xi\neq\eta$ or $\sigma\neq\tau$. Lepage (1971) showed that the asymptotic distribution of L, under the null hypothesis, is χ^2 with $\nu=2$ degrees of freedom.

There are several other tests that are very similar to the Lepage test. It is possible to define a test using Wilcoxon and Mood scores instead of the Wilcoxon and Ansari–Bradley scores. A test that has similar properties to those of the Wilcoxon–Mood test was suggested by O'Brien (1988). In O'Brien's test the indicator variable is used as the dependent variable, the rank as one independent variable, and the squared rank as the other independent variable in a multiple regression model. O'Gorman (1995) showed that the power of O'Brien's test closely approximates the power of the Wilcoxon–Mood test. We should note that special software may be needed to perform the Wilcoxon–Ansari–Bradley test and the Wilcoxon–Mood test, but O'Brien's test can be easily computed using any multiple regression program.

8.2.3 An Adaptive Test for Location and Scale

Büning and Thadewald (2000) proposed an adaptive test of location and scale. Their idea was to use a measure of tailweight

$$\hat{T}=\frac{\hat{\xi}_{.975}-\hat{\xi}_{.025}}{\hat{\xi}_{.875}-\hat{\xi}_{.125}},$$

where $\hat{\xi}_p$ is the $100p$th percentile of the combined sample, to select an appropriate set of scores for a test of location and scale. The tests selected have the same general form as the Lepage test but use different rank scores. If $\hat{T}\leq 1.5$, the adaptive method determines that it is a short-tailed distribution so it selects rank scores designed for short-tailed distribution for the location scores and the absolute values of these scores for the scale scores. If $1.5<\hat{T}\leq 1.9$, the adaptive method determines that the tails have moderate length so it selects normal scores for the location scores and Klotz scores for the scale scores. If $\hat{T}>1.9$, the adaptive method determines that it is a long-tailed distribution so it selects rank scores that are suited to long-tailed distributions for the location scores and Mood scores for the scale scores. For a description of the rank scores used in the component tests, the reader should consult Büning and Thadewald (2000).

Since the selection statistic $\hat{T}$ is a function of the order statistics of the combined samples and the scores are based on the ranks of the combined samples, the resulting test will maintain its level of significance. Büning and Thadewald (2000) showed, using simulation studies with $n_1=n_2=25$, that their adaptive test had greater power than the Lepage test for six symmetric distributions with several location and scale parameters.

8.2.4 An Adaptive Test for a General Alternative

If the data in the first sample were generated from $F(y)$ and the data in the second sample were generated from $G(y)$, then a test of $H_0: F(y)=G(y)$ for all y, against the general alternative $H_a: F(y)\neq G(y)$ for at least one y, would be appropriate if we felt that the distributions might not be stochastically ordered. The traditional tests for these general alternatives are the Kolmogorov–Smirnov test and the Cramér–von Mises test. An adaptive test for the general alternative has been published by Büning (2002). In this test measures of skewness and tailweight are used to select appropriate tests of the Kolmogorov–Smirnov type and the Cramér–von Mises type. Büning (2002) shows, via a simulation study, that the adaptive test has high power compared to the other tests for general alternatives.

8.3 Using Adaptive Methods with Discretized Data

In our simulations we have always used continuous distributions to evaluate the performance of the adaptive tests and confidence intervals. However, when we obtain data from an experimental or an observational study we often find that they have been discretized into a few values. For example, if we are recording the height of adults, we might record the height to the nearest inch. Consequently, in a data set we may have only 10 or 15 distinct levels of the dependent variable, which will produce many tied values. Will the adaptive methods described in the previous chapters work with many tied values?

In order to investigate this question, we used discrete distributions obtained by rounding the continuous variates. In these simulations we rounded the response to the nearest multiple of 0.5 so that the response would be the nearest value in the set $\{\ldots, -1.0, -0.5, 0.0, 0.5, 1.0, \ldots\}$. This process of discretization is most easily seen in Figures 8.5 and 8.6, which show the approximate normal distribution and the discretized distribution. It can be seen that, by rounding the standard normal variates to the nearest multiple of 0.5, most of the observations take on one of nine values. For skewed and long-tailed distributions the discretized variates may have more than nine values, but there may be a higher proportion of tied values.

We performed simulations to evaluate the performance of the two-sample adaptive test in the same manner as in Chapter 2, except we used the discretized generalized lambda distributions with rounding to the nearest multiple of 0.5. For the simulations with sample sizes of $n_1 = n_2 = 10$, $n_1 = n_2 = 20$, and $n_1 = 48$, $n_2 = 12$, we obtained empirical sizes between 4.37% and 5.21% for the adaptive test, using $\alpha = .05$, with the nine error distributions. Consequently, the simulation results for the size, which are not tabulated, show that the adaptive test maintained its level of significance.

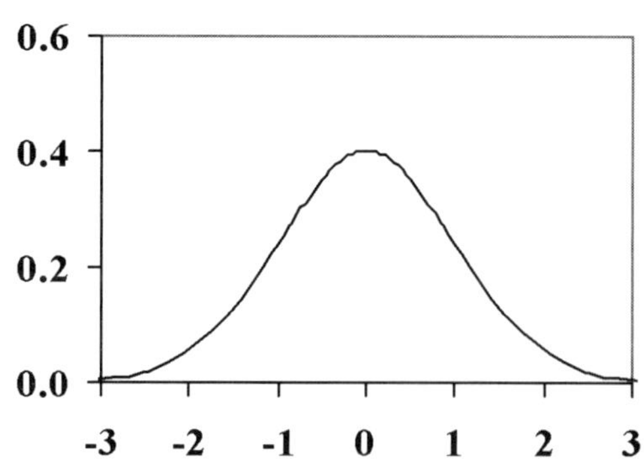

Figure 8.5. *The approximate normal distribution.*

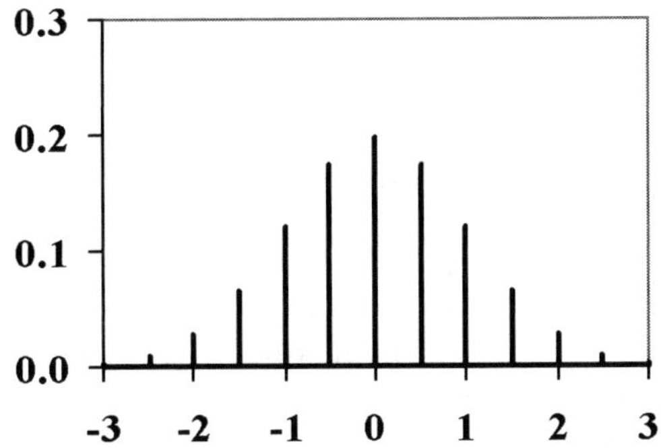

Figure 8.6. *The discrete distribution obtained by discretizing the approximate normal distribution.*

The empirical power of the t test and adaptive test are shown in Table 8.2. These results show that the adaptive test has greater power than the t test, even when the variates have been discretized to the extent that many values are tied. In addition, the results show that when the error distribution was approximately normal, the power of the adaptive test was within 2% of the power of the t test. Overall, it appears that this amount of rounding had little effect on the relative performance of the two tests.

Table 8.2. *The empirical power in percent for the proposed test and the pooled t test for data rounded to the nearest multiple of one-half, using a significance level of* $\alpha = .05$.

	Distributions								
Skewness	0	0	0	1	1	1	2	2	2
Kurtosis	1.8	3.0	6.0	4.2	5.4	8.4	11.4	12.6	15.6
				$n_1 = 10,\ n_2 = 10$					
Adaptive	49.7	48.4	53.2	55.9	54.9	57.1	65.9	65.7	64.0
t test	48.2	50.1	51.8	51.9	51.8	53.8	55.2	56.9	57.1
				$n_1 = 20,\ n_2 = 20$					
Adaptive	55.9	48.5	54.0	60.0	57.1	57.6	71.3	69.9	65.8
t test	45.2	49.8	51.0	50.2	50.8	51.7	53.0	53.9	53.0
				$n_1 = 48,\ n_2 = 12$					
Adaptive	58.7	47.7	54.7	63.2	58.6	59.0	74.7	71.8	68.5
t test	48.6	48.9	50.5	50.4	50.7	52.7	54.3	54.3	54.8

8.4 An Adaptive Multiple Comparison Procedure

8.4.1 Simulation Results

After experimental data are obtained from a one-way layout, researchers often need to compare several of the group means by performing tests of significance of the null hypothesis

$H_0 : \mu_i = \mu_j$, where μ_i and μ_j are the population means for the ith and jth levels. If several comparisons are made, each using a significance level of α, the chance that one of the null hypotheses will be rejected will exceed α. In order to keep the experimentwise error rate at α, a multiple comparison procedure is often used. Hill, Padmanabhan, and Puri (1988) proposed two adaptive schemes for multiple comparisons. In this section we investigate the properties of an adaptive multiple comparison procedure that uses the testing methods developed in Chapters 2 to 4.

In order to maintain the experimentwise significance level near α we will use a Bonferroni multiple comparison procedure. In the Bonferroni method we perform the tests of $H_0 : \mu_i = \mu_j$ in the usual manner, but we use an adjusted significance level of $\alpha' = \alpha/g$ for each test, where g is the number of comparisons. In order to compare the ith level of a factor to the jth level of that factor in a one-way layout, we will test $H_0 : \mu_i = \mu_j$ against $H_a : \mu_i \neq \mu_j$. We will carry out a test of $H_0 : \mu_i = \mu_j$ by comparing the sum of squared errors of a full model to the sum of squared errors of a reduced model. The full model includes an indicator variable for each level of the factor. The reduced model includes $K - 1$ indicator variables, with one of the variables coded to indicate that it is an observation from either the ith or the jth level.

A simulation study was used to determine if this adaptive procedure would maintain its size and have a power advantage over the traditional Bonferroni procedure. We used a one-way layout with $K = 4$ treatment groups with ten observations in each group. Our goal was to test the three null hypotheses $H_0 : \mu_1 = \mu_2$, $H_0 : \mu_1 = \mu_3$, and $H_0 : \mu_1 = \mu_4$. If we use the Bonferroni multiple comparison procedure with $\alpha = .05$ and $g = 3$, we reject the null hypothesis if $p < .05/3 = .0167$. In our simulation study we used 10,000 data sets in each of nine simulations corresponding to the nine error distributions. The empirical significance levels of the adaptive test and the t test are shown in Table 8.3. These results indicate that we rejected $H_0 : \mu_1 = \mu_4$ in 1.34% to 1.78% of these data sets, which shows that the adaptive test maintained its level of significance. The empirical powers of this test, which are also shown in Table 8.3, demonstrate that the adaptive test is more powerful than the traditional t test for nonnormal errors and is nearly as powerful for normal errors when the Bonferroni approach is used. Similar results were found for the tests of $H_0 : \mu_1 = \mu_2$ and $H_0 : \mu_1 = \mu_3$.

Table 8.3. *Empirical power estimates in percent for the multiple comparison procedure for the test of* $H_0 : \mu_1 = \mu_4$ *against* $H_a : \mu_1 \neq \mu_4$ *with* $n_1 = n_2 = n_3 = n_4 = 10$ *and* $\alpha' = .05/3 = .0167$.

	Distributions								
Skewness	0	0	0	1	1	1	2	2	2
Kurtosis	1.8	3.0	6.0	4.2	5.4	8.4	11.4	12.6	15.6
	Empirical Significance Level								
Adaptive	1.54	1.74	1.34	1.59	1.74	1.77	1.69	1.75	1.78
t test	1.49	1.65	1.40	1.50	1.67	1.69	1.60	1.52	1.71
	Empirical Power								
Adaptive	36.0	33.0	38.2	42.7	41.5	43.5	55.3	53.6	51.1
t test	32.0	34.1	35.6	35.2	36.1	38.3	38.8	40.1	39.0

These simulation results were not too surprising since the individual tests in a Bonferroni multiple comparison procedure are special cases of the general tests described in Chapter 3. Thus, we would expect that the individual tests would have the overall characteristics of the tests described in Chapters 2, 3, and 4.

8.4.2 A SAS Macro for Multiple Comparison Procedures

An adaptive multiple comparison procedure can be performed by using the SAS macro "awlsmc." This macro uses the same adaptive procedure that the "awls" macro uses but is designed to be used only when the X_A matrix is created from an indicator variable for one treatment, which must be specified as a classification variable. For example, consider the data of Cawson et al. (1974) in Table 8.4, which concerns the cortisol level in three groups of pregnant women.

Table 8.4. *The cortisol level in three groups of pregnant women.*

Group	Cortisol Level	Mean
1	154, 211, 262, 287, 304, 307, 323, 339, 356, 454	300
2	355, 362, 455, 465, 468, 501	434
3	207, 343, 687, 772, 838, 1048	649

The traditional F test for a one-way layout produced a p-value of $p = .0039$ and the adaptive test gave a p-value of $p = .0145$, so it is reasonable to conclude that there are differences between the groups. To perform the adaptive pairwise comparisons we used the "awlsmc" macro, which was designed to give the p-values for all pairwise comparisons. This example had "level" as the dependent variable and "group" as the classification variable on the data set "cortisol." The macro calls used to perform the overall analysis and the multiple comparisons are

```
%awls(dataset=cortisol, y=level, xr=, nvarsred=0,
      xa=group, classa=group, seed=978673);
%awlsmc(y=level, xa=group, classa=group, seed=492833);
```

This macro will output p-values for all pairwise comparisons. For the test of $H_0 : \mu_1 = \mu_2$ the traditional t test yields $p = .152$ while the adaptive test gives $p = .033$. For this comparison the adaptive result appears to be more reasonable because all of the observations in group 2 exceeded most of the observations in group 1. For the test of $H_0 : \mu_1 = \mu_3$ the traditional t test yields $p = .001$ while the adaptive test gives $p = .01$. To compare group 2 to group 3 the traditional t test yields $p = .046$ while the adaptive test gives $p = .172$. Again, the adaptive test result seems more reasonable because two of the values in group 3 are less than all of the values in group 2, while the other four values in group 3 are greater than all of the values in group 2. Admittedly, this data set has much more variability in group 3 than in group 2, but the adaptive procedure took this into account by downweighting some of the observations in group 3.

8.5 Concluding Remarks

Real data are not easy to analyze. Even if a moderate amount of data was collected, it is often not clear what statistical model would be most appropriate. We may find that the data are not continuous but have been discretized when they were recorded, or we may find that the treatment group has greater variability than the control group. Consequently, before we use statistical procedures we would like to know how sensitive the procedures are to departures from the assumptions.

Fortunately, statistical tests, whether adaptive or nonadaptive, are not greatly affected by a moderate amount of discretization. We found, by inspecting the simulation results in Table 8.2, that the pooled t test and the adaptive test had high power to detect differences even with a moderate amount of discretization.

We also found in section 8.2.1 that two-sample tests maintain their size and have reasonable power even when the treatment effect is somewhat variable. These results are in agreement with those found by O'Gorman (1997b), who used several nonconstant treatment effects with traditional parametric, nonparametric, and rank-based adaptive tests. Although the tests are reasonably robust to departures from a shift model, it should be emphasized that they may not be robust if the treatment effect is such that the distributions are not stochastically ordered.

If a researcher has good reason to believe that the treatment could greatly increase the variability, then it appears that a test for location would be inappropriate. Fortunately, good tests for location and scale have been published. The tests of Lepage (1971) and O'Brien (1988) should be considered. For even more power the adaptive test by Büning and Thadewald (2000) could be used.

Researchers who analyze experimental data often need to use some form of multiple comparison procedure in the analysis. The multiple comparison procedures are designed to maintain the experimentwise error rate at a predetermined level. The SAS macro described in section 8.4 can be used to perform the Bonferroni multiple comparison test by generating a p-value from the adaptive test for every possible pairwise comparison. The simulation results shown in Table 8.2 indicate that these tests will have the same properties as the adaptive tests described in Chapters 2, 3, and 4.

The rank-based adaptive tests described in section 8.1 and Hogg's adaptive two-sample test described in section 1.3 often have higher power than the traditional tests. These are important tests because they maintain their size for all error distributions. The chief difficulty of the rank-based tests is that they are limited to a few fairly simple models. In contrast, the adaptive WLS tests are applicable to a variety of models. They are robust for validity and efficiency and can be used for testing any subset of parameters in a linear model.

Appendix A

Macro 1: SAS Macro for Adaptive Testing of Subsets of Regression Coefficients

```
/*
  This file contains a SAS macro for performing the adaptive weighted
  least squares test for a subset of a model.

  Author:  Tom O'Gorman
           Northern Illinois University
           DeKalb, Illinois 60115

           e-mail: ogorman@math.niu.edu

  This program was carefully checked on January 16, 2003 and I believe it
  to be correct.  However, the author is not responsible for any errors
  that may still be lurking in the code.

  Notes:

     1)  The default number of permutations is set to 2000.  It can be
         set to a larger or smaller value.

     2)  No missing values are allowed in the dependent or independent
         variables.

     3)  If variables are formatted then the formats will need to be placed
         before the macro.

     4)  The number of variables in the Xr matrix, not including the
         intercept, must be specified in the macro call as the variable
         nvarsred=.  (This is necessary so that the correct number of
         degrees of freedom will be used in the t distribution.)

     5)  The number of observations must not exceed 10000.

*/

 options nocenter linesize=80  nonotes ;

%macro awls(dataset=, y=, xr=, classr=, nvarsred=, xa=, classa=,
            seed=0, nperm=2000, short=0);
```

```
data awlsdata(keep= &y &xr &xa) dxa(keep=&xa );
   set &dataset;
     output awlsdata;
     output dxa;
proc glm data=awlsdata   %if &short ne 0 %then noprint;;
  classes  &classr &classa;
  model &y=&xr &xa ;
 %if &classa ne %then lsmeans &classa/stderr tdiff pdiff; ;
  title2 ' traditional analysis -- no adaptation';

proc glm data=awlsdata ;
  classes  &classr ;
  model &y=&xr/p;
  output out=dresid rstudent=di p=predicted r=residual;
  title2 'adaptive weighted least squares analysis';

data dweights(keep=di dcdf weight saswt );
  array dr{*} d1-d10000;
  array smcdf{*} f1-f10000;
  array points{*} p1 p2 p3;
  array smpoint{*} smp1 smp2 smp3;
  set dresid nobs=rn end=last;
  retain h d1-d10000 f1-f10000 ;
  if _n_ = 1  then  h=1.587/rn**0.333333;
  i=_n_;
  dr{i}=di;
  if last then do;
    do j=1 to rn;
      sum=0.0;
      do k=1 to rn;
        z=(dr{j}-dr{k})/h;
        sum=sum+probnorm(z);
        end;
      smcdf{j}=sum/rn;
      end;
    p1=-2;
    p2=0;
    p3=2;
    do j = 1 to 3;
      sum = 0.0;
      do k = 1 to rn;
        z = (points{j}-dr{k})/h;
        sum = sum + probnorm(z);
        end;
      smpoint{j}=sum/rn;
      end;
 nextp: if smp2 lt 0.5 then do;
          p1=p2;
          smp1=smp2;
          end;
        if smp2 gt 0.5 then do;
          p3=p2;
          smp3=smp2;
          end;
        if smp2 eq 0.5 then go to over;
        p2=(p1+p3)/2;
        sum=0.0;
```

```
        do k=1 to rn;
          z=(p2-dr{k})/h;
          sum=sum+probnorm(z);
          end;
        smp2=sum/rn;
        if (p3-p1) lt 0.00001 then go to over;
                                  else go to nextp;
 over:  dmedian=p2;
        df = rn - %eval(&nvarsred) - 2;
        put dmedian=;
        do k=1 to rn;
          dc = dr{k} - dmedian;
          di=dr{k};
          dcdf=smcdf{k};
          weight=tinv(dcdf,df)/dc;
          saswt=weight*weight;
          output;
          end;
   end;
data dresid(keep=predicted residual);
  set dresid;
data dall;
  merge awlsdata dweights dresid;
  %if &short eq 0 %then %do;
    proc print data=dall ;
     var &y &xr &xa predicted residual di weight saswt ;
    title2' Data with adaptive weights    SASWT=weights used by SAS';
    %end;
proc glm data=dall outstat=dstar noprint; ;
  classes  &classr &classa;
  model &y=&xr &xa/ss1 ;
  weight saswt;
data dssstar(keep=ssstar count);
  set dstar;
  if _source_ eq 'ERROR' then do;
    ssstar=ss;
    count=0;
    output;
    end;
data dseed;
   nextseed=&seed;

  %do i=1 %to &nperm;
     data dperm(drop=nextseed) dseed(keep=nextseed);
       retain seed1 ;
       set dseed(in=inseed) dxa(in=indxa) end=last;
       if inseed then seed1=nextseed;
       if indxa then do;
         call ranuni(seed1,rnd);
         output dperm;
         end;
       if last then do;
         nextseed=seed1;
         output dseed;
         end;
     proc sort data=dperm; by rnd;
     data dxrxa ;
```

```
        merge dall dperm;
      proc glm data=dxrxa outstat=dssperm noprint;
        classes  &classr &classa;
        model &y=&xr &xa/ss1 ;
        weight saswt;
      data dssperm;
        set dssperm;
        if _source_='ERROR' then do;
          ssperm=ss;
          output;
          end;
      data dssstar(keep=ssstar count);
        merge dssstar dssperm;
        if ssperm le ssstar then count=count+1;
      %end;
      data _null_;
        set dssstar;
        p_value=(count+1)/(&nperm+1);
   if &short eq 0 then do;
     file print ;
     put '1';
     put ;
     put '  ******************************************************* ';
     put ;
     put "       Adaptive weighted least squares analysis";
     put ;
     put " Number of permutations = &nperm with first seed = &seed" ;
     put " SAS data set:   &dataset ";
     put ;
     put " Dependent variable                           :  &y ";
     put ;
     put " Independent variables in reduced model  : Intercept  &xr " ;
     put "       class variables in reduced model   :  &classr ";
     put;
     put " Independent variables in complete model : Intercept &xr  &xa ";
     put "       class variables in complete model : &classr  &classa";
     put;
     put '  p-value for adaptive test = ' p_value 8.4 ;
     put;
     put '  ******************************************************* ';
     put;
     end;
    else do;
     file print noprint;
     put "            with &nperm permutations,   p-value = " p_value 8.4 ;
     put;
     end;
run;
%mend;
```

Macro 2: SAS Macro for an Adaptive Test for Paired Data

```
/*
This file contains a SAS macro that performs an adaptive test for
paired samples.
```

```
      Author:  Tom O'Gorman
               Northern Illinois University
               DeKalb, Illinois 60115

               e-mail: ogorman@math.niu.edu

      This program was carefully checked on January 16, 2003 and I believe it
      to be correct.  However, the author is not responsible for any errors
      that may still be lurking in the code.

      Notes:

         1)  The default number of permutations is set to 2000.  It can be
             set to a larger or smaller value.

         2)  If variables are formatted then the formats will need to be placed
             before the macro.

         3)  The number of observations must not exceed 5000.
    */

 options nocenter linesize=80  nonotes ;

%macro oneawls(dataset=, y=, seed=0, nperm=2000);
data awlsdata;
  set &dataset;
  if nmiss( of &y) eq 0 then output;
proc univariate data=awlsdata;
  var &y;
  title2 ' traditional analysis -- no adaptation';
data dy2;
  set awlsdata;
  output;
  &y=-&y;
  output;
  title2 ' Adaptive test for paired data';
proc univariate data=dy2 pctldef=4 noprint;
  var &y;
  output out=dstat qrange=iqr;
data dzdata2;
  set dstat(in=instat) awlsdata;
  retain sd;
  if instat then do;
    sd=iqr/1.349;
    end;
   else do;
    z=&y/sd;
    output;
    z=-z;
    output;
    end;
data dsmooth (keep=rawdata zz zcdf zinv wt) dt(keep=t0 count);
  array dy{*} ay1-ay10000;
  array dz{*} az1-az10000;
  array smcdf{*} af1-af10000;
  set dzdata2 nobs=rn end=last;
  retain h ay1-ay10000 az1-az10000 af1-af10000;
```

```
   if _n_ =1 then do;
     h=.5/(rn/2)**.33333;
     put h=;
     end;
   i=_n_;
   dy{i}=&y;
   dz{i}=z;
   if last then do;
     do j=1 to rn;
       sum=0.0;
       do k=1 to rn;
         zdist=(dz{j}-dz{k})/h;
         sum=sum + probnorm(zdist);
         end;
       smcdf{j}=sum/rn;
       end;
     t0=0;
     do k=1 to rn-1 by 2;
       rawdata=dy{k};
       zz=dz{k};
       zcdf=smcdf{k};
       zinv=probit(zcdf);
       wt=zinv/zz;
       t0 = t0 + wt*wt*dy{k};
       output dsmooth;
       end;
     count=0;
     output dt;
     end;
proc print data=dsmooth;

data dseed;
   set dt;
   nextseed=&seed;

  %do i=1 %to &nperm;
     data dseed(keep=t0 count nextseed) ;
     retain t0r seedr countr tstar;
       set dseed(in=inseed) dsmooth(in=insmooth) end=last;
       if inseed then do;
         t0r=t0;
         countr=count;
         seedr=nextseed;
         tstar=0.0;
         end;
       if insmooth then do;
         call ranuni(seedr,rnd);
         if rnd le 0.5 then sign=-1;
                        else sign=+1;
         tstar = tstar + wt*wt*sign*rawdata;
         end;
       if last then do;
          t0=t0r;
          if tstar gt t0r then countr = countr + 1;
          count  = countr;
          nextseed = seedr;
          output dseed;
```

```
        end;
  %end;
     data _null_;
       set dseed;
       p_value=(count+1)/(&nperm+1);
       tp_value=2*p_value;
    file print ;
    put '1';
    put ;
    put '  ******************************************************** ';
    put ;
    put "       Adaptive weighted least squares analysis";
    put ;
    put " Number of permutations = &nperm with first seed = &seed" ;
    put " SAS data set:   &dataset ";
    put ;
    put " Variable name                          :  &y ";
    put ;
    put '  one-tailed p-value for adaptive test = ' p_value 8.4 ;
    put '  two-tailed p-value for adaptive test = ' tp_value 8.4;
    put;
    put '  ******************************************************** ';
    put;
%mend;
```

Macro 3: SAS Macro for Calculating the Confidence Interval for the Mean of a Symmetric Population

```
    /*
      This file contains a SAS macro for calculating the 95% confidence
      interval for the mean of a symmetric distribution.

      Author:  Tom O'Gorman
               Northern Illinois University
               DeKalb, Illinois 60115

               e-mail: ogorman@math.niu.edu

      This program was carefully checked on January 17, 2003 and I believe it
      to be correct.   However, the author is not responsible for any errors
      that may still be lurking in the code.


      Notes:

        1)  The number of observations must not exceed 1000.

        2)  If the variables are formatted then the formats will need to be
            placed before the macro.

        3)  The macro will update the estimates of the lower and upper limits
            until the number of iterations equals maxiter, which is set to
            8000 by default.

    */
```

```
 options nocenter linesize=80 nonotes ;

%macro cione(dataset=, y=,  seed=0, maxiter=8000);
data dseed;
  nextseed=&seed;
data dupdate;
  c=0;
data dgarth(keep=&y );
  set &dataset;
  if &y ne .  then output;
data;
  set dgarth nobs=rn;
  if rn gt 1000 then do;
    file print noprint notitle;
    put ' The number of observations exceeds 1000;  Job cancelled';
    abort;
    end;
proc means data=dgarth alpha=.05 n mean median std lclm uclm;
  var &y;
  output out=dtheta median=thetahat lclm=li uclm=ui;
  title1 '  Analysis with 95% Confidence Intervals -- No Adaptation';
data dupdate;
  set dtheta;
  c=li;
  iter=1;
  title1 ' Adaptive 95% Confidence Intervals';
  file print noprint notitle;
  put ' Adaptive Weighted Least Squares Confidence Intervals';
  put ' The search method was proposed by ';
  put ' Garthwaite, P. H. , Biometrics, vol. 52, 1387-1393, 1996.';
  put ' ';
  put ' ';
  put '  **************************************************';
  put '         now begin adaptive lower limit';
  put '  **************************************************';
  put ' ';

%do i=1 %to &maxiter;

%oneci( y=&y,  printout=0);

data dupdate;
  merge dupdate dseed;
  if tperm lt tunperm then li=li+17.46*abs(thetahat-li)*0.025/(iter+22);
                      else li=li-17.46*abs(thetahat-li)*0.975/(iter+22);
  file print noprint notitle;
  if iter le 10 or (iter le 1000 and iter eq 100*(int(iter/100)))
     or iter eq 1000*(int(iter/1000)) then
          put ' after iteration ' iter ' adaptive lower limit = ' li 12.3;
  iter=iter+1;
  c=li;
%end;


data dupdate;
  set dupdate;
  file print noprint notitle;
```

```
  put ' ';
  put '   ************************************************';
  put '          now begin adaptive upper limit';
  put '   ************************************************';
  put ' ';
  c=ui;
  iter=1;
%do i=1 %to &maxiter;

%oneci( y=&y, printout=0);

data dupdate;
  merge dupdate dseed;
  if tperm ge tunperm then ui=ui-17.46*abs(ui-thetahat)*0.025/(iter+22);
                      else ui=ui+17.46*abs(ui-thetahat)*0.975/(iter+22);
  file print noprint notitle;
  if iter le 10 or (iter le 1000 and iter eq 100*(int(iter/100)))
     or iter eq 1000*(int(iter/1000)) then
        put ' after iteration ' iter ' adaptive upper limit = ' ui 12.3;
  iter=iter+1;
  c=ui;
%end;

%mend;


%macro oneci( y=,  printout=2);

data awlsdata(keep= &y yadj ) ;
   set dupdate(in=inup)  dgarth(in=indata);
   retain const;
   if inup then const=c;
   if indata then do;
     yadj=&y-const;
     output awlsdata;
     end;
%if &printout eq 2 %then %do;
   proc print data=awlsdata;
   %end;
data dy2;
  set awlsdata;
  output;
  yadj=-yadj;
  output;
proc univariate data=dy2 pctldef=4 noprint;
  var yadj;
  output out=dstat qrange=iqr;
data dz2;
  set dstat(in=instat) dy2;
  retain sd;
  if instat then do;
    sd=iqr/1.349;
    end;
   else do;
    z=yadj/sd;
    output;
    end;
```

```
data dsmooth (keep= adjdata zz zcdf wt) dt(keep=tunperm);
  array dy{*} ay1-ay2000;
  array dz{*} az1-az2000;
  array smcdf{*} af1-af2000;
  set dz2 nobs=rn end=last;
  retain h ay1-ay2000 az1-az2000 af1-af2000;
  if _n_ =1 then do;
    h=1.26/(rn/2)**.33333;
    end;
  i=_n_;
  dy{i}=yadj;
  dz{i}=z;
  if last then do;
    do j=1 to rn;
      sum=0.0;
      do k=1 to rn;
        zdist=(dz{j}-dz{k})/h;
        sum=sum + probnorm(zdist);
        end;
      smcdf{j}=sum/rn;
      end;
    tunperm=0;
    do k=1 to rn-1 by 2;
      adjdata=dy{k};
      zz=dz{k};
      zcdf=smcdf{k};
      wt=probit(zcdf)/zz;
      tunperm = tunperm + wt*wt*adjdata;
      output dsmooth;
      end;
    output dt;
    end;
 data dseed(keep=tunperm tperm nextseed) ;
     retain tunpermr seedr  tpermr;
       set dt(in=int) dseed(in=inseed) dsmooth(in=insmooth) end=last;
       if int then do;
         tunpermr=tunperm;
         tpermr=0.0;
         end;
       if inseed then do;
         seedr=nextseed;
         end;
       if insmooth then do;
         call ranuni(seedr,rnd);
         if rnd le 0.5 then sign=-1;
                      else sign=+1;
         tpermr = tpermr + wt*wt*sign*adjdata;
         end;
       if last then do;
          tperm=tpermr;
          tunperm=tunpermr;
          nextseed = seedr;
          output dseed;
          end;

%mend;
;
```

Macro 4: SAS Macro for Calculating the Confidence Interval for a Parameter in a Linear Model

```
/*
  This file contains a SAS macro for calculating the 95% confidence
  interval for one regression parameter in a multiple regression model.

  Author:   Tom O'Gorman
            Northern Illinois University
            DeKalb, Illinois 60115

            e-mail: ogorman@math.niu.edu

  This program was carefully checked on January 17, 2003 and I believe it
  to be correct.  However, the author is not responsible for any errors
  that may still be lurking in the code.

  Important note: This program requires version 8.0 or above.
                  If a lower version of SAS is used remove the "clb"
                  option in the model statement on the first
                  proc reg.

  Other notes:

     1)  The number of observations must not exceed 10000.

     2)  If the variables are formatted then the formats will need to be
         placed before the macro.

     3)  The macro will update the estimates of the lower and upper limits
         until the number of iterations equals maxiter, which is set to
         8000 by default.

 */

 options nocenter linesize=80  nonotes;

%macro ciawls(dataset=, y=, xr=, xa=, seed=0, maxiter=8000);
data dseed;
  nextseed=&seed;
data dupdate;
  c=0;
data dgarth(keep=&y &xr &xa);
  set &dataset;
  if nmiss( of &y &xr &xa) eq 0 then output;
proc reg data=dgarth;
  model &y=&xr &xa/clb;
title1 ' Regression Analysis with 95% Confidence Intervals -- No Adaptation';
%awlsci( y=&y, xr=&xr, xa=&xa, printout=0);
data dthetaha(keep=thetahat);
  set dtunperm;
  thetahat=beta;
file print noprint;
title1 ' Adaptive Weighted Least Squares Confidence Intervals';
title2 ' Search method based on method proposed by ';
title3 ' Garthwaite, P. H. , Biometrics, vol. 52, 1387-1393, 1996.';
 put ;
```

```
proc reg data=dgarth outest=dtheta outseb noprint; ;
  model &y=&xr &xa;
data dupdate;
  set dtunperm(in=inunperm) dtheta;
  retain thetahat theta;
  if inunperm  then thetahat=beta;
  if _type_ eq 'PARMS' then theta=&xa;
  if _type_ eq 'SEB'   then do;
  file print noprint notitle;
  put ' ';
  put '  **************************************************';
  put '          now begin adaptive lower limit';
  put '  **************************************************';
  put ' ';

                              ui=theta+1.96*&xa;
                              li=theta-1.96*&xa;
                              c=li;
                              iter=1;
                              output;
                              end;
%do i=1 %to &maxiter;

%awlsci( y=&y, xr=&xr, xa=&xa, printout=0);

data dupdate;
  merge dupdate dtunperm dtperm;
  if tperm lt tunperm then li=li+17.46*abs(thetahat-li)*0.025/(iter+22);
                      else li=li-17.46*abs(thetahat-li)*0.975/(iter+22);
  file print noprint notitle;
  if iter le 10 or (iter le 1000 and iter eq 100*(int(iter/100)))
     or iter eq 1000*(int(iter/1000)) then
        put ' after iteration ' iter ' adaptive lower limit = ' li 12.3;
  iter=iter+1;
  c=li;
%end;


data dupdate;
  set dupdate;
  file print noprint notitle;
  put ' ';
  put '  **************************************************';
  put '          now begin adaptive upper limit';
  put '  **************************************************';
  put ' ';
  c=ui;
  iter=1;
%do i=1 %to &maxiter;

%awlsci( y=&y, xr=&xr, xa=&xa, printout=0);

data dupdate;
  merge dupdate dtunperm dtperm;
  if tperm ge tunperm then ui=ui-17.46*abs(ui-thetahat)*0.025/(iter+22);
                      else ui=ui+17.46*abs(ui-thetahat)*0.975/(iter+22);
  file print noprint notitle;
```

```
  if iter le 10 or (iter le 1000 and iter eq 100*(int(iter/100)))
     or iter eq 1000*(int(iter/1000)) then
          put ' after iteration ' iter ' adaptive upper limit = ' ui 12.3;
  iter=iter+1;
  c=ui;
%end;

%mend;


%macro awlsci( y=, xr=, xa=, printout=2);

data awlsdata(keep= &y yadj &xr &xa) dxa(keep=&xa );
   set dupdate(in=inup)  dgarth(in=indata);
   retain const;
   if inup then const=c;
   if indata then do;
     yadj=&y-const*&xa;
     output awlsdata;
     output dxa;
     end;
%if &printout eq 2 %then %do;
   proc print data=awlsdata;
   proc print data=dxa;
   %end;

proc glm data=awlsdata noprint;
  model yadj=&xr;
  output out=dresid rstudent=di;
  title2 'adaptive weighted least squares analysis';

data dweights(keep=di dcdf weight saswt);
  array dr{*} d1-d10000;
  array smcdf{*} f1-f10000;
  array points{*} p1 p2 p3;
  array smpoint{*} smp1 smp2 smp3;
  set dresid nobs=rn end=last;
  retain h d1-d10000 f1-f10000 ;
  if _n_ = 1  then  h=1.587/rn**0.333333;
  i=_n_;
  dr{i}=di;
  if last then do;
    do j=1 to rn;
      sum=0.0;
      do k=1 to rn;
        z=(dr{j}-dr{k})/h;
        sum=sum+probnorm(z);
        end;
      smcdf{j}=sum/rn;
      end;
    p1=-2;
    p2=0;
    p3=2;
    do j = 1 to 3;
      sum = 0.0;
      do k = 1 to rn;
        z = (points{j}-dr{k})/h;
```

```
        sum = sum + probnorm(z);
        end;
      smpoint{j}=sum/rn;
      end;
 nextp: if smp2 lt 0.5 then do;
          p1=p2;
          smp1=smp2;
          end;
        if smp2 gt 0.5 then do;
          p3=p2;
          smp3=smp2;
          end;
        if smp2 eq 0.5 then go to over;
        p2=(p1+p3)/2;
        sum=0.0;
        do k=1 to rn;
          z=(p2-dr{k})/h;
          sum=sum+probnorm(z);
          end;
        smp2=sum/rn;
        if (p3-p1) lt 0.00001 then go to over;
                                else go to nextp;
 over:  dmedian=p2;
        df=rn-n(of &xr)-2;
        do k=1 to rn;
          dc = dr{k} - dmedian;
          di=dr{k};
          dcdf=smcdf{k};
          weight=tinv(dcdf,df)/dc;
          saswt=weight*weight;
          output;
          end;
   end;

data dall;
  merge awlsdata dweights;
  %if &printout eq 2 %then %do;
    proc print data=dall ;
     var &y yadj &xr &xa weight saswt;
    title2' Data, after adjustment     SASWT=weights used by SAS';
      %end;

proc reg data=dall outest=dunperm outseb %if &printout eq 0 %then noprint; ;
  model yadj = &xr &xa;
  weight saswt;
data dtunperm(keep=tunperm beta);
  set dunperm;
  retain beta;
  if _type_ eq 'PARMS' then beta=&xa;
  if _type_ eq 'SEB'   then do;
    tunperm=beta/&xa;
    output;
    end;

data dperm(drop=nextseed) dseed(keep=nextseed);
       retain seed1 ;
       set dseed(in=inseed) dxa(in=indxa) end=last;
```

```
      if inseed then seed1=nextseed;
      if indxa then do;
        call ranuni(seed1,rnd);
        output dperm;
        end;
      if last then do;
        nextseed=seed1;
        output dseed;
        end;
    proc sort data=dperm; by rnd;
    data dxrxa ;
      merge dall dperm;
    proc reg data=dxrxa outest=dperm outseb %if &printout eq 0 %then noprint;;
      model yadj = &xr &xa ;
      weight saswt;
    data dtperm(keep=tperm);
      set dperm;
      retain beta;
      if _type_ eq 'PARMS' then beta=&xa;
      if _type_ eq 'SEB'   then do;
        tperm=beta/&xa;
        output;
        end;
%mend;
;
```

Macro 5: SAS Macro for Adaptive Estimation of the Mean of a Symmetric Distribution

```
    /*
      This file contains a SAS macro that performs an adaptive estimate
      of the mean of a symmetric distribution.

      Author:  Tom O'Gorman
               Northern Illinois University
               DeKalb, Illinois 60115

               e-mail: ogorman@math.niu.edu

      This program was carefully checked on January 17, 2003 and I believe it
      to be correct.  However, the author is not responsible for any errors
      that may still be lurking in the code.

      Notes:

       1) The number of observations must not exceed 5000.

       2) If variables are formatted then the formats will need to be placed
          before the macro.

   */

 options nocenter linesize=80  ;

%macro estone(dataset=, y=);
data awlsdata;
```

```
  set &dataset;
  if nmiss( of &y) eq 0 then output;
proc univariate data=awlsdata pctldef=4;
  var &y;
  output out=dstat median=median;
  title2 ' traditional analysis -- no adaptation';
data dcenter2;
  set dstat(in=instat) awlsdata;
  retain ymedian;
  if instat then ymedian=median;
           else do;
             ycenter=&y-ymedian;
             output;
             ycenter=-ycenter;
             output;
             end;
proc univariate data=dcenter2 pctldef=4 noprint;
  var ycenter;
  output out=dstat qrange=iqr;
data dzdata2;
  set dstat(in=instat) dcenter2;
  retain sd;
  if instat then do;
    sd=iqr/1.349;
    end;
   else do;
    z=ycenter/sd;
    output;
    end;
data dsmooth (keep=rawdata zz zcdf wt) dbetahat(keep=betahat);
  array dy{*} ay1-ay10000;
  array dz{*} az1-az10000;
  array smcdf{*} af1-af10000;
  set dzdata2 nobs=rn end=last;
  retain h ay1-ay10000 az1-az10000 af1-af10000;
  if _n_ =1 then do;
    h=1.26/(rn/2)**.33333;
    put h=;
    end;
  i=_n_;
  dy{i}=&y;
  dz{i}=z;
  if last then do;
    do j=1 to rn;
      sum=0.0;
      do k=1 to rn;
        zdist=(dz{j}-dz{k})/h;
        sum=sum + probnorm(zdist);
        end;
      smcdf{j}=sum/rn;
      end;
    sumwwy=0;
    sumww=0;
    do k=1 to rn-1 by 2;
      rawdata=dy{k};
      zz=dz{k};
      zcdf=smcdf{k};
```

```
      if abs(zz) le 0.00001 then wt=1.0;
                                 else wt=probit(zcdf)/zz;
      sumwwy=sumwwy+ wt*wt*dy{k};
      sumww=sumww+wt*wt;
      output dsmooth;
      end;
    betahat=sumwwy/sumww;
    output dbetahat;
    end;
proc print data=dsmooth;
data ;
  set dbetahat;
  file print;
    put ;
    put '   ******************************************************* ';
    put ;
    put "       Adaptive Estimate of the Mean";
    put ;
    put "       Variable name              :  &y ";
    put ;
    put '       Estimate = ' betahat 8.4 ;
    put;
    put '   ******************************************************* ';
    put;
%mend;
```

Macro 6: SAS Macro for Adaptive Estimation of Regression Parameters

```
    /*
      This file contains a SAS macro for calculating the point
      estimates for regression parameters in a multiple regression model.

      Author:  Tom O'Gorman
               Northern Illinois University
               DeKalb, Illinois 60115

               e-mail: ogorman@math.niu.edu

      This program was carefully checked on January 17, 2003 and I believe it
      to be correct.  However, the author is not responsible for any errors
      that may still be lurking in the code.

      Notes:

        1)  The number of observations must not exceed 10000.

        2)  If the variables are formatted then the formats will need to be
            placed before the macro.

   */

 options nocenter linesize=80  ;

%macro pointest(dataset=, y=, x=,  printout=1);
```

```
data dreg;
  set &dataset;
  if nmiss(of &y &x) eq 0 then output;

proc reg data=dreg %if &printout eq 0 %then noprint; ;
  model &y=&x/influence;
  output out=dresid rstudent=di;
  title2 'Traditional ordinary least squares estimates';

data dweights(keep=di dcdf weight saswt);
  array dr{*} d1-d10000;
  array smcdf{*} f1-f10000;
  array points{*} p1 p2 p3;
  array smpoint{*} smp1 smp2 smp3;
  set dresid nobs=rn end=last;
  retain h d1-d10000 f1-f10000 ;
  if _n_ = 1  then  h=1.587/rn**0.333333;
  i=_n_;
  dr{i}=di;
  if last then do;
    do j=1 to rn;
      sum=0.0;
      do k=1 to rn;
        z=(dr{j}-dr{k})/h;
        sum=sum+probnorm(z);
        end;
      smcdf{j}=sum/rn;
      end;
    p1=-2;
    p2=0;
    p3=2;
    do j = 1 to 3;
      sum = 0.0;
      do k = 1 to rn;
        z = (points{j}-dr{k})/h;
        sum = sum + probnorm(z);
        end;
      smpoint{j}=sum/rn;
      end;
nextp: if smp2 lt 0.5 then do;
         p1=p2;
         smp1=smp2;
         end;
       if smp2 gt 0.5 then do;
         p3=p2;
         smp3=smp2;
         end;
       if smp2 eq 0.5 then go to over;
       p2=(p1+p3)/2;
       sum=0.0;
       do k=1 to rn;
         z=(p2-dr{k})/h;
         sum=sum+probnorm(z);
         end;
       smp2=sum/rn;
       if (p3-p1) lt 0.00001 then go to over;
                            else go to nextp;
```

```
 over:  dmedian=p2;
        df = rn - n(of &x) - 2;
      put dmedian= df=;
        do k=1 to rn;
          dc = dr{k} - dmedian;
          di=dr{k};
          dcdf=smcdf{k};
          weight=tinv(dcdf,df)/dc;
          saswt=weight*weight;
      put k= dc= di= dcdf=;
          output;
          end;
   end;

data dall;
  merge &dataset dweights;
  %if &printout eq 1 %then %do;
    proc print data=dall ;
     var &y  &x  weight saswt;
    title2' Data, after adjustment     SASWT=weights used by SAS';
      %end;

proc reg data=dall outest=dslope noprint;
  model &y = &x;
  weight saswt;
    title2 ' Adaptive Point Estimates for Parameters ';
proc print data=dslope;
  var intercept &x;

%mend;
```

Macro 7: SAS Macro for Performing Adaptive Multiple Comparisons

```
    /*
      This file contains a SAS macro for performing the
      Bonferroni multiple comparison procedure with the
      adaptive weighted least squares test. It must be used
      with the awls macro.

      Author:  Tom O'Gorman
               Northern Illinois University
               DeKalb, Illinois 60115

               e-mail: ogorman@math.niu.edu

      This program was carefully checked on December 20, 2002 and I believe
      it to be correct.   However, the author is not responsible for any
      errors that may still be present in the code.

      Notes:

        1)  This macro must be included in a SAS program that also has
            the "awls" macro.
```

```
        2)  No missing values are allowed in the dependent or independent
            variables.

        3)  If variables are formatted then the formats will need to be placed
            before the macro.

        4)  The macro is to be used with a one-way layout.  Do not specify
            anything for xr or classr; they are not used as inputs in
            this macro.

        5)  The indicator variable for the groups should be specified as
            the xa and classa variables.

        6)  The number of independent variables should not be specified; it
            will be calculated by this macro before the call to the awls
            macro.

        7)  The number of observations must not exceed 10000.

  */

%macro awlsmc(dataset=,y=,xa=,classa=,seed=0, nperm=2000);
 proc sort data=&dataset out=dsorted;
   by &classa;
 data dk;
   retain k 0 ;
   keep k ;
   set dsorted end=last;
     by &classa;
   if first.&classa then do;
     k=k+1;
     call symput ('km',k);
     km2=k-2;
     call symput ('km2m',km2);
     end;
   if last then output;
   run;
 %do imc=2 %to &km;
 %do jmc=1 %to &imc-1;
  data expand;
    length lower $ 8;
    retain lower upper;
    retain c 0;
    file print noprint;
    keep  &y &xa newclass ia;
    set  dsorted end=last;
      by &classa;
    if first.&classa then do;
        c = c + 1;
        if c = &jmc then lower=&classa;
        end;
    if c = &imc then do;
                                newclass=lower;
                                ia=1;
                                upper=&classa;
                                end;
```

```
                        else do;
                             newclass=&classa;
                             ia=0;
                             end;
    if last then put "  compare " lower " to " upper " for variable &classa";
  %awls(dataset=expand,y=&y, xr=newclass, classr=newclass, nvarsred=&km2m,
        xa=ia, nperm=&nperm, seed=&seed, short=1);
  %end;
  %end;
%mend;
;
```

Appendix B. Data Sets

Table B.1. *Cigarette sales by state in* 1970.

State	Sales	Price	Income	Age	Percent Female
AL	89.8	42.7	2948.0	27.0	51.7
AK	121.3	41.8	4644.0	22.9	45.7
AZ	115.2	38.5	3665.0	26.3	50.8
AR	100.3	38.8	2878.0	29.1	51.5
CA	123.0	39.7	4493.0	28.1	50.8
CO	124.8	31.1	3855.0	26.2	50.7
CT	120.0	45.5	4917.0	29.1	51.5
DE	155.0	41.3	4524.0	26.8	51.3
DC	200.4	32.6	5079.0	28.4	53.5
FL	123.6	43.8	3738.0	32.3	51.8
GA	109.9	35.8	3354.0	25.9	51.4
HI	82.1	36.7	4623.0	25.0	48.0
ID	102.4	33.6	3290.0	26.4	50.1
IL	124.8	41.4	4507.0	28.6	51.5
IN	134.6	32.2	3772.0	27.2	51.3
IO	108.5	38.5	3751.0	28.8	51.4
KA	114.0	38.9	3853.0	28.7	51.0
KY	155.8	30.1	3112.0	27.5	50.9
LA	115.9	39.3	3090.0	24.8	51.4

Table B.2. *Cigarette sales by state in* 1970 *(continued).*

State	Sales	Price	Income	Age	Percent Female
ME	128.5	38.8	3302.0	28.0	51.3
MD	123.5	34.2	4309.0	27.1	51.1
MA	124.3	41.0	4340.0	29.0	52.2
MI	128.6	39.2	4180.0	26.3	51.0
MN	104.3	40.1	3859.0	26.8	51.0
MS	93.4	37.5	2626.0	25.1	51.6
MO	121.3	36.8	3781.0	29.4	51.8
MT	111.2	34.7	3500.0	27.1	50.0
NB	108.1	34.7	3789.0	28.6	51.2
NV	189.5	44.0	4563.0	27.8	49.3
NH	265.7	34.1	3737.0	28.0	51.1
NJ	120.7	41.7	4701.0	30.1	51.6
NM	90.0	41.7	3077.0	23.9	50.7
NY	119.0	41.7	4712.0	30.3	52.2
NC	172.4	29.4	3252.0	26.5	51.0
ND	93.8	38.9	3086.0	26.4	49.5
OH	121.6	38.1	4020.0	27.7	51.5
OK	108.4	39.8	3387.0	29.4	51.3
OR	157.0	29.0	3719.0	29.0	51.0
PA	107.3	44.7	3971.0	30.7	52.0
RI	123.9	40.2	3959.0	29.2	50.9
SC	103.6	34.3	2990.0	24.8	50.9
SD	92.7	38.5	3123.0	27.4	50.3
TN	99.8	41.6	3119.0	28.1	51.6
TX	106.4	42.0	3606.0	26.4	51.0
UT	65.5	36.6	3227.0	23.1	50.6
VT	122.6	39.5	3468.0	26.8	51.1
VA	124.3	30.2	3712.0	26.8	50.6
WA	96.7	40.3	4053.0	27.5	50.3
WV	114.5	41.6	3061.0	30.0	51.6
WI	106.4	40.2	3812.0	27.2	50.9
WY	132.2	34.4	3815.0	27.2	50.0

Bibliography

ACT (2002). Web site: *www.act.org*, Iowa City, Iowa.

Anderson, J. W., Story, L., Sieling, B., Chen, W. L., Petro, M. S., and Story, J. (1984). Hypocholesterolemic effects of oat-bran or bean intake for hypercholesterolemic men. *The American Journal of Clinical Nutrition*, 40, 1146–1155.

Anderson, M. J. and Legendre, P. (1999). An empirical comparison of permutation methods for tests of partial regression coefficients in a linear model. *Journal of Statistical Computation and Simulation*, 62, 271–303.

Andrews, D. F., Bickel, P. J., Hampel, F. R., Huber, P. J., Rogers, W. H., and Tukey, J. W. (1972). *Robust Estimates of Location*, Princeton University Press, Princeton, New Jersey.

Atkinson, A. C. (1986). Comment: aspects of diagnostic regression analysis (discussion of paper by Chatterjee and Hadi), *Statistical Science,* 1, 397–402.

Atkinson, A. C. (1988). Transformations unmasked, *Technometrics,* 30, 311–318.

Behnen, K. and Neuhaus, G. (1989). *Rank Tests with Estimated Scores and Their Application*, Teubner, Stuttgart, Germany.

Belsley, D. A., Kuh, E., and Welsch, R. E. (1980). *Regression Diagnostics: Identifying Influential Data and Sources of Collinearity*, Wiley, New York.

Bickel, P. J. and Doksum, K. A. (1977). *Mathematical Statistics: Basic Ideas and Selected Topics*, Holden-Day, Oakland, California.

Blair, R. C. and Higgins, J. J. (1985). Comparison of the power of the paired samples t test to that of Wilcoxon's signed-ranks test under various population shapes, *Psychological Bulletin*, 97, 119–128.

Blumberg, E. M., West, P. M., and Ellis, F. W. (1954). A possible relationship between psychological factors and human cancers, *Psychosomatic Medicine*, 16, 277–286.

Büning, H. (1996). Adaptive tests for the c-sample location problem—the case of two-sided alternatives. *Communications in Statistics—Theory and Methods,* 25, 1569–1582.

Büning, H. (1999). Adaptive Jonckheere-type tests for ordered alternatives. *Journal of Applied Statistics*, 26, 541–551.

Büning, H. (2002). An adaptive distribution-free test for the general two-sample problem. *Computational Statistics*, 17, 297–313.

Büning, H. and Kössler, W. (1998). Adaptive tests for umbrella alternatives. *Biometrical Journal*, 40, 573–587.

Büning, H. and Thadewald, T. (2000). An adaptive two-sample location-scale test of Lepage type for symmetric distributions. *Journal of Statistical Computation and Simulation*, 65, 287–310.

Cawson, M. J., Anderson, A. B., Turnbull, A. C., and Lampe, L. (1974). Cortisol, cortisone, and 11-deoxycortisol levels in human umbilical and maternal plasma in relation to the onset of labour. *Journal of Obstetric Gynaecology British Commonwealth*, 1974, 737–745.

Chatterjee, S., Hadi, A. S., and Price, B. (2000). *Regression Analysis by Example*, 3rd ed., Wiley, New York.

Clarke, B. R. (1994). Empirical evidence for adaptive confidence intervals and identification of outliers using methods of trimming. *Australian Journal of Statistics*, 36, 45–58.

Coleman, J. S., Cambell, E. Q., Hobson, C. J., McPartland, J., Mood, A. M., Weinfield, F. D., and York, R. L. (1966). *Equality of Educational Opportunity*, two volumes, Office of Education, U.S. Department of Health, Washington, D.C.

Conover, W. J. (1973). Rank tests for one sample, two samples and k samples without the assumption of a continuous distribution function. *Annals of Statistics*, 1, 1105–1125.

Davis, C. S. and Chung, Y. (1995). Randomization model methods for evaluating treatment efficacy in multicenter clinical trials. *Biometrics*, 51, 1163–1174.

Davison, A. C. and Hinkley, D. V. (1997). *Bootstrap Methods and Their Application,* Cambridge University Press, Cambridge, U.K.

Dodge, Y. and Jurečková, J. (2000). *Adaptive Linear Regression,* Springer-Verlag, New York.

Farber, S. L. (1981). *Identical Twins Reared Apart*, Basic Books, New York.

Fenstad, G. U., Kjaernes, M., and Walløe, L. (1980). Robust estimation of standard deviation. *Journal of Statistical Computation and Simulation*, 10, 113–132.

Fleiss, J. L. (1986). *The Design and Analysis of Clinical Experiments*, Wiley, New York.

Garthwaite, P. (1996). Confidence intervals from randomization tests. *Biometrics*, 52, 1387–1393.

Garthwaite, P. H. and Buckland, S. T. (1992). Generating Monte Carlo confidence intervals by the Robbins–Monro process. *Applied Statistics*, 41, 159–171.

Gastwirth, J. L. (1965). Percentile modifications of two sample rank tests. *Journal of the American Statistical Association*, 60, 1127–1141.

Haith, D. A. (1976). Land use and water quality in New York rivers. *Journal of the Environmental Engineering Division, Proceedings of the American Society of Civil Engineers*, 102, 1–15.

Hampel, F. R., Ronchetti, E. M., Rousseeuw, P. J., and Stahel, W. A. (1986). *Robust Statistics,* Wiley, New York.

Hájek, J. (1962). Asymptotically most powerful rank-order tests. *Annals of Mathematical Statistics*, 33, 1124–1147.

Hall, P. and Padmanabhan, A. (1997). Adaptive inference for the two-sample scale problem. *Technometrics*, 39, 412–422.

Han, C. and Hawkins, D. L. (1994). A smooth adaptive estimator of the mean of a symmetric or asymmetric distribution. *Communications in Statistics—Theory and Methods*, 23, 1–10.

Hettmansperger, T. P. and Sheather, S. J. (1992). A cautionary note on the method of least median squares. *The American Statistician*, 46, 79–83.

Hill, N. J. and Padmanabhan, A. R. (1991). Some adaptive robust estimators which work with real data. *Biometrical Journal*, 33, 81–101.

Hill, N. J., Padmanabhan, A. R., and Puri, M. L. (1988). Adaptive nonparametric procedures and applications. *Applied Statistics*, 37, 205–218.

Hogg, R. V., Fisher, D. M., and Randles, R. H. (1975). A two-sample adaptive distribution-free test. *Journal of the American Statistical Association*, 70, 656–661.

Hogg, R. V. and Lenth, R. V. (1984). A review of some adaptive statistical techniques. *Communications in Statistics—Theory and Methods*, 13, 1551–1579.

Hollander, M. and Wolfe, D. A. (1973). *Nonparametric Statistical Methods*, Wiley, New York.

Huber, P. J. (1973). Robust regression: asymptotics, conjectures and Monte Carlo. *Annals of Statistics*, 1, 799–821.

Hüsler, J. (1987). On the two-sample adaptive distribution free test. *Communications in Statistics—Theory and Methods*, 16, 55–68.

Jiang, C. L. and Hunt, J. N. (1983). The relation between freely chosen meals and body habitus. *The American Journal of Clinical Nutrition*, 38, 32–40.

Johnson, M. E. (1987). *Multivariate Statistical Simulation,* Wiley, New York.

Kennedy, P. E. and Cade, B. S. (1996). Randomization tests for multiple regression. *Communications in Statistics—Simulation and Computation*, 25, 923–936.

Kleinbaum, D. G., Kupper, L. L., Muller, K. E., and Nizam, A. (1998). *Applied Regression Analysis and Multivariable Methods*, International Thomson Publishing, Pacific Grove, California.

Lemmer, H. H. (1993). Adaptive tests for the median. *IEEE Transactions on Reliability*, 42, 442–448.

Lehmann, E. L. (1975). *Nonparametrics: Statistical Methods Based on Ranks*, Holden-Day, San Francisco.

Lepage, Y. (1971). A combination of Wilcoxon's and Ansari–Bradley's statistics. *Biometrika*, 58, 213–217.

Linton, L. R., Edgington, E. S., and Davies, R. W. (1989). A view of niche overlap amenable to statistical analysis. *Canadian Journal of Zoology*, 67, 55–60.

Liu, Q. and Chi, G. Y. H. (2001). On sample size and inference for two-stage adaptive designs. *Biometrics*, 57, 172–177.

Lo, S. and Han, C. (1997). Robust linear regression using smooth adaptive estimators. *Communications in Statistics—Simulation and Computation,* 26, 1–19.

Mack, G. A. and Wolfe, D. A. (1981). K-sample rank tests for umbrella alternatives. *Journal of the American Statistical Association*, 76, 175–181.

Manly, B. F. J. (1997). *Randomization, Bootstrap and Monte Carlo Methods in Biology,* Chapman and Hall, London.

Marazzi, A. (1993). *Algorithms, Routines, and S Functions for Robust Statistics*, Wadsworth, Pacific Grove, California.

McKean, J. W. and Sheather, S. J. (1991). Small sample properties of robust analyses of linear models based on R-estimates: A survey, in *Directions in Robust Statistics and Diagnostics: Part* II, W. Stahel and S. Weisberg, eds., Springer-Verlag, New York.

Mitchell, H. H., Burroughs, W., and Beadles, J. R. (1936). The significance and accuracy of biological values of proteins computed from nitrogen metabolism data. *Journal of Nutrition*, 11, 257–274.

Mosteller, F. and Tukey, J. W. (1977). *Data Analysis and Regression.* Addison-Wesley, Reading, Massachusetts.

Neter, J., Wasserman, W., and Kutner, M. H. (1989). *Applied Regression Models*, 2nd ed., Irwin, Burr Ridge, Illinois.

Newcomb. S. (1891). Measures of the velocity of light made under the direction of the Secretary of the Navy during the years 1880–1882. *Astronomical Papers*, 2, 107–230. U.S. Nautical Almanac Office.

O'Brien, P. C. (1988). Comparing two samples: extensions of the t, rank-sum, and log-rank tests. *Journal of the American Statistical Association*, 83, 52–61.

O'Gorman, T. W. (1995). The effect of unequal variances on the power of several two-sample tests. *Communications in Statistics—Simulation and Computation*, 24, 853–867.

O'Gorman, T. W. (1996). An adaptive two-sample test based on modified Wilcoxon scores. *Communications in Statistics—Simulation and Computation*, 25, 459–479.

O'Gorman, T. W. (1997a). An adaptive test for the one-way layout. *The Canadian Journal of Statistics*, 25, 269–279.

O'Gorman, T. W. (1997b). A comparison of an adaptive two-sample test to the t-test and the rank-sum test. *Communications in Statistics—Simulation and Computation*, 26, 1393–1411.

O'Gorman, T. W. (2001a). An adaptive permutation test procedure for several common tests of significance. *Computational Statistics and Data Analysis*, 35, 335–350.

O'Gorman, T. W. (2001b). Adaptive estimation using weighted least squares. *The Australian and New Zealand Journal of Statistics*, 43, 287–297.

O'Gorman, T. W. (2001c). Using adaptive weighted least squares to reduce the lengths of confidence intervals. *The Canadian Journal of Statistics*, 29, 459–471.

O'Gorman, T. W. (2002). An adaptive test of significance for a subset of regression coefficients. *Statistics in Medicine*, 21, 3527–3542.

Pitman, E. J. G. (1937). Significance tests which may be applied to samples from any population. *Journal of the Royal Statistical Society*, B4, 119–130.

Polansky, A. M. (1998). A smooth nonparametric approach to process capability. *Quality and Reliability Engineering International*, 14, 43–48.

Policello, G. E. and Hettmansperger, T. P. (1976). Adaptive robust procedures for the one-sample location problem. *Journal of the American Statistical Association*, 71, 624–633.

Portnoy, S. and Koenker, R. (1989). Adaptive L-estimation for linear models. *Annals of Statistics*, 17, 362–381.

Powell, G. L. and Russell, A. P. (1984). The diet of the eastern short-horned lizard (Phrynosoma douglassi brevirostre) in Alberta and its relationship to sexual size dimorphism. *Canadian Journal of Zoology*, 62, 428–440.

Ramberg, J. S., Dudewicz, E. J., Tadikamalla, P. R., and Mykytka, E. F. (1979). A probability distribution and its uses in fitting data. *Technometrics*, 21, 201–214.

Randles, R. H. and Hogg, R. V. (1973). Adaptive distribution-free tests. *Communications in Statistics*, 2, 337–356.

Randles, R. H. and Wolfe, D. A. (1979). *Introduction to the Theory of Nonparametric Statistics*, Wiley, New York.

Reed, J. F. and Stark, D. B. (1994). Robustness estimators of location: A classification of linear and adaptive estimators. *Journal of Applied Statistics*, 21, 95–124.

Robbins, H. and Monro, S. (1951). A stochastic approximation method. *Annals of Mathematical Statistics*, 22, 400–407.

Rousseeuw, P. J. (1984). Least median of squares regression. *Journal of the American Statistical Association*, 79, 871–880.

Rousseeuw, P. J. and Leroy, A. M. (1987). *Robust Regression and Outlier Detection*, Wiley, New York.

Ruberg, S. J. (1986). A continuously adaptive nonparametric two-sample test. *Communications in Statistics—Theory and Methods*, 15, 2899–2920.

Ryan, T. P. (1997). *Modern Regression Methods*, Wiley, New York.

Simon, D. (2001). *A Nutritional Assessment of a Mature Adult Population in Santo Domingo, Dominican Republic.* Master's Thesis, Northern Illinois University. Raw data courtesy of the author.

Simpson, D. G., Ruppert, D., and Carroll, R. J. (1992). On one-step GM estimates and stability of inferences in linear regression. *Journal of the American Statistical Association*, 87, 439–450.

Simpson, G. G., Roe, A., and Lewontin, R. C. (1960). *Quantitative Zoology*, Harcourt, Brace and World, New York.

Snedecor, G. W. and Cochran, W. G. (1967). *Statistical Methods.* Iowa State University Press, Ames, Iowa.

Stigler, S. M. (1977). Do robust estimators work with real data?, *The Annals of Statistics*, 5, 1055–1098.

Stromberg, A. J. (1993). Comment on "A cautionary note on the method of least median squares," *The American Statistician*, 47, 87.

Stromberg, A. J. (1992). Computing the Exact Least Median of Squares Estimate and Stability Diagnostics in Multiple Linear Regression, Technical Report 329, University of Kentucky, Department of Statistics, Lexington, Kentucky.

Yuh, L. and Hogg, R. V. (1988). On adaptive M-Regression. *Biometrics*, 44, 433–445.

Index